I0482652

Disclaimer

The publisher of this book is by no way associated with the National Institute of Standards and Technology (NIST). The NIST did not publish this book. It was published by 50 page publications under the public domain license.

50 Page Publications.

Book Title: Modeling and Measuring the Effects of Portable Gasoline Powered Generator Exhaust on Indoor Carbon Monoxide Level

Book Author: Steven J. Emmerich; Andrew K. Persily; Wang Liangzhu;

Book Abstract: The U.S. Consumer Product Safety Commission (CPSC) is concerned about the hazard of acute residential carbon monoxide (CO) exposures from portable gasoline powered generators that can result in death or serious and/or lasting adverse health effects in exposed individuals. As an initial approach to characterizing these hazards, CPSC measured the emissions from generators by testing them in a small test chamber (Brown 2006). CPSC subsequently contracted with the University of Alabama (UA) to develop and construct low CO-emission prototype generators using off-the-shelf technologies installed on commercially-available portable generators. Under an interagency agreement with CPSC, NIST conducted a series of tests to characterize the indoor time course profiles of CO concentrations resulting from portable generators operating in the attached garage of a home under various use and environmental conditions, to evaluate the performance of low CO-emission prototype generators, and to provide model validation data. The data was also used as input to a simulation analysis conducted to examine the potential performance of the low CO-emission prototypes under a wider range of operating conditions.

Citation: NIST TN - 1781

Keywords: carbon monoxide; CONTAM; exposure; indoor air quality; health; measurements; multizone airflow model; simulation

NIST Technical Note 1781

Modeling and Measuring the Effects of Portable Gasoline Powered Generator Exhaust on Indoor Carbon Monoxide Level

Steven J Emmerich
Andrew K Persily
Liangzhu (Leon) Wang

http://dx.doi.org/10.6028/NIST.TN.1781

National Institute of
Standards and Technology
U.S. Department of Commerce

NIST Technical Note 1781

Modeling and Measuring the Effects of Portable Gasoline Powered Generator Exhaust on Indoor Carbon Monoxide Level

Steven J Emmerich
Andrew K Persily
Energy and Environment Division
Engineering Laboratory

Liangzhu (Leon) Wang
Concordia University

http://dx.doi.org/10.6028/NIST.TN.1781

February 2013

U.S. Department of Commerce
Rebecca Blank, Acting Secretary

National Institute of Standards and Technology
Patrick D. Gallagher, Under Secretary of Commerce for Standards and Technology and Director

Certain commercial entities, equipment, or materials may be identified in this document in order to describe an experimental procedure or concept adequately. Such identification is not intended to imply recommendation or endorsement by the National Institute of Standards and Technology, nor is it intended to imply that the entities, materials, or equipment are necessarily the best available for the purpose.

National Institute of Standards and Technology Technical Note 1781
Natl. Inst. Stand. Technol. Tech. Note 1781, 141 pages (February 2013)
http://dx.doi.org/10.6028/NIST.TN.1781
CODEN: NTNOEF

Abstract

The U.S. Consumer Product Safety Commission (CPSC) is concerned about the hazard of acute residential carbon monoxide (CO) exposures from portable gasoline powered generators that can result in death or serious and/or lasting adverse health effects in exposed individuals. In an initial approach to characterizing these hazards, CPSC measured the CO emission rates from generators by testing them in a small, laboratory test chamber (Brown 2006). CPSC subsequently contracted with the University of Alabama (UA) to develop a low CO-emission prototype generator by adapting off-the-shelf emission control technologies on a commercially-available portable generator and then to construct multiple units for testing. Under an interagency agreement with CPSC, NIST conducted a series of tests to characterize the indoor CO concentrations resulting from portable generators operating in the attached garage of a house under various use and environmental conditions so CPSC staff could analyze the safety implications of operating generators under these conditions. The tested generators include both unmodified and modified low CO emission prototype configurations. NIST used those test data to validate the ability of the CONTAM indoor air quality (IAQ) model to predict CO levels in the garage and the house and to develop an estimate of the uncertainty of these predictions relative to measured values. NIST also conducted tests with the generators operating in a one-zone shed to derive their CO emission and O_2 consumption rates. These rates were used both as inputs to the model validation effort as well as in simulation analyses conducted to examine the potential performance of the low CO-emission prototype under a wider range of operating conditions.

To determine generator CO emission rates under more realistic conditions than those attained in a small, laboratory test chamber, NIST conducted tests on the unmodified and modified generators (i.e. without and with CO emission controls) in a single-zone shed. For two different unmodified generators (i.e., without CO emission controls), it was found that CO emissions ranged from a low of around 500 g/h at near ambient O_2 levels to a high of nearly 4000 g/h as O_2 approached 17 %. The rates of CO generation and O_2 consumption in these unmodified generators were affected by multiple parameters, with the O_2 level in the space and the actual electrical output of the generator being two of the most important. Tests performed below 17 % O_2 showed a drop off in CO emissions due to poor engine performance under these conditions. Tests of two modified, prototype generators (i.e., with CO emission controls) showed CO emissions reductions of over 90 % depending on the specific emission controls and operating conditions and no trend toward higher emission rates was seen as O_2 levels dropped to 18 %.

A series of tests were also conducted to measure the emission and transport of CO when operating portable gasoline-powered generators in an attached garage. This series of tests included both unmodified and prototype generators operated in the garage attached to NIST's manufactured test house. Testing was conducted under seven different test house/garage configurations to evaluate their impacts on the buildup of CO in the garage and its transport into different rooms in the house. The configurations studied included two different garage bay door positions (fully closed or open 0.6 m), two connecting door settings between the garage and the family room (fully closed or open 5 cm), and two house central heating, ventilating, and air conditioning (HVAC) fan settings (on or off). CO concentrations varied widely with peak house CO concentrations ranging from under 10 μL/L to over 10,000 μL/L (note that μL/L are equivalent to ppm$_v$). As expected, the highest concentrations in the house resulted from operation of the unmodified generator in the garage with the bay door closed and the house access door open 5 cm. The lowest concentrations resulted from operation of a modified low CO-emission prototype in the garage with the bay door open 0.6 m and the house access door closed. These garage tests documented reductions of 85 % to 98 % in CO concentrations in the house due to emissions from two modified, prototype low CO-emission portable generators compared to a "stock" generator. Note that these results apply to the specific units tested and that other units, modifications and test conditions may produce different results.

An extensive model validation effort using the multizone airflow and IAQ model CONTAM was carried out using the data from the seven tests that were conducted with a generator operating in the attached garage of the test house to compare predicted CO concentrations with measured values. The agreement between the measurements and predictions of the O_2 concentrations in the garage and of the average CO calculated for the house zones was excellent for the collective set of data; all of the calculated statistical values met the ASTM D5157 criteria for comparing IAQ model predictions and measurements. The agreement, however, was somewhat worse for the garage CO concentrations, with some parameters falling slightly outside the ASTM criteria limits. Overall, the average individual house zone and garage CO concentration predictions and measurements were within about 20 % and 30 % respectively when averaged over all cases.

Forty-two simulations were then performed with the NIST CONTAM model to examine the potential performance of the prototype generator under a wider range of conditions than studied during the experiments at the test house. All of the simulations were based on the NIST manufactured test house and the tested low CO emission prototype generator with and without a catalyst integrated in the muffler (referred to as catmuffler and noncat muffler, respectively). Model parameters that were varied included ambient conditions, CO emission rates, source locations and door positions. The highest house CO concentrations were found for the generator with the noncat muffler operated in the utility room of the house, with indoor CO concentrations reaching 2000 µL/L for some cases. Operation of the generator with the cat muffler substantially reduced CO concentrations, however, they still reached levels of 280 µL/L to 600 µL/L for cases with the generator located in the utility room. The lowest indoor CO concentrations resulted from operation of the generator with cat muffler in the garage with CO concentrations in the house reaching 10 µL/L to 160 µL/L. Simulations also showed that, as expected, closed bedroom doors resulted in less uniform indoor concentrations and higher peak indoor zone concentrations, though the impact varied greatly from about a 10 % to about a 100 % increase.

Keywords

Generator; carbon monoxide; CONTAM; exposure; indoor air quality; health; measurements; multizone airflow model; simulation

Table of Contents

List of Figures

List of Tables

Nomenclature

A_{in}	hourly air change rate of a test space evaluated at the ambient temperature, h^{-1}
A_{out}	hourly air change rate of a test space evaluated at the space temperature, h^{-1}
C_{CO}	CO concentration, mg/m^3
C_{H_2O,t_1}	water vapor volumetric concentration at time t_1, m^3/m^3
C_{H_2O,t_2}	water vapor volumetric concentration at time t_2, m^3/m^3
$C_{O_2}^*$	non-dimensional O_2 concentration
C_{out}	gas volumetric concentration outside the space, m^3/m^3
$C_{SF_6,in}$	SF_6 concentration in the intake of the generator, m^3/m^3
$C_{SF_6,ex}$	SF_6 concentration in the exhaust of the generator, m^3/m^3
D_{gen}	displacement of the generator engine, m^3
ELA	Effective leakage area, cm^2 at 4 Pa
$K_{m,\rho}$	$\rho_{m,o} / \rho_{tm,i}$, ratio of density outside and inside the test space of a gas component "m"
$K_{SF_6,d}$	SF_6 decomposition ratio in the generator
L_o	output load of a generator
L_o^*	non-dimensional generator load
P	partial pressure of water vapor, Pa
P_{sat}	saturated water vapor pressure, Pa
$Q_{gen,in}$	volumetric flow rate through the generator at the intake gas temperature, m^3/h
Q_{in}	volumetric air inflow of a test space, m^3/h
Q_{out}	volumetric air outflow of a test space, m^3/h
$Q_{gen,ex}$	volumetric flow rate through the generator at the exhaust gas temperature, m^3/h
RPM	revolutions per minute of the generator engine, min^{-1}
S_C	source emission rate of a gas "C" at the air temperature of the test space, kg/h
S_{CO}^*	non-dimensional CO emission rate
$S_{CO,min}$	minimum CO emission rate, g/h
$S_{CO,max}$	maximum CO emission rate, g/h
S_{H_2O}	water emission rate from the generator, kg/h
S_{gen}	mass added from the generator to the test space, kg/h
$S_{O_2}^*$	non-dimensional O_2 consumption rate
T	gas temperature, K
$T_{gen,in}$	temperature of the generator intake gas, K
$T_{gen,ex}$	temperature of the generator exhaust gas, K
V_s	volume of the test space, m^3
ϕ	water vapor relative humidity
η_{gen}	volumetric efficiency of the generator engine
$\rho_{C,in}$	density of a gas component inside the test space, kg/m^3
$\rho_{C,out}$	density of a gas component outside the space, kg/m^3
ρ_{in}	gas mixture density in the test space, kg/m^3
$\rho_{m,out}$	density of gas mixture ($= \rho_{air}$) outside the test space, kg/m^3

Introduction

Background

The U.S. Consumer Product Safety Commission (CPSC) is concerned about the hazard of acute residential carbon monoxide (CO) exposures from portable gasoline powered generators that can result in death or serious and/or lasting adverse health effects in exposed individuals. As of April 2012, CPSC databases contain records of at least 755 (695 from generator use alone, 60 from generator use in conjunction with another CO-producing consumer product) deaths from CO poisoning associated with consumer use of generators in the period of 1999 through 2011 (Hnatov 2012). In addition, the percentage of estimated non-fire, consumer product-related CO poisoning deaths specifically associated with generators for CPSC's four most recent years of data are 51 % (2005), 49 % (2006), 38 % (2007), and 49 % (2008) (Hnatov 2011). Typically, these deaths occur when consumers use a generator in an enclosed or partially enclosed space or outdoors near an open door, window or vent, and they often occur after severe weather events such as hurricanes and ice or snow storms. The initial health impact of CO is caused by anoxia: deprivation of oxygen supply. When inhaled, CO preferentially binds with the oxygen carrier in the red blood cells, hemoglobin (Hb), to form carboxyhemoglobin (COHb), which causes the anoxia (Stewart 1975).

Since the possession of household generators in the U.S. has climbed continuously in recent years, from an estimated 9.2 million units in 2002 to 10.6 million units in 2005, CPSC is working to avoid future generator-related CO poisoning incidents, especially those associated with operating a generator indoors (CPSC 2006). Measures have been taken to educate people not to operate generators indoors and to require manufacturers of portable generators to warn consumers of CO hazards with a warning label (CFR 2007). In order to understand the CO exposures associated with such incidents and their potential reduction, the emission characteristics of these generators need to be better characterized.

In an initial approach to characterizing these hazards, CPSC measured the CO emission rates from generators by testing them in a small, laboratory test chamber (Brown 2006). Using those data, CPSC performed preliminary indoor air quality (IAQ) modeling and estimated that a 92 % reduction in the CO emission rate based on these measurements would likely result in a significant delay and reduced severity of the CO exposure in areas of a home remote from the generator location (Inkster 2006). CPSC subsequently worked with the University of Alabama (UA) to develop a low CO-emission prototype generator by adapting off-the-shelf emission control technologies on a commercially-available portable generator. UA then constructed multiple prototype generators by adapting the same emission control strategy onto other units powered by the same model engine. In conjunction with these efforts, CPSC established an interagency agreement with the National Institute of Standards and Technology (NIST) to model indoor time course profiles of CO concentrations throughout a single-family residence resulting from CO emissions from a portable generator operating in the attached garage of a home under various use and environmental conditions to enable CSPC staff to understand the safety implications of operating a portable generator in these conditions. The agreement was later expanded to include a series of tests to provide empirical data to further characterize the hazard by measuring the emission and transport of CO when generators are operated in an actual building, to test them in a one-zone shed to derive their CO emission and O_2 consumption rates, and to provide model validation data.

1

An interim report by Emmerich (2011) presents data from a series of tests of both unmodified and UA-modified low CO emission prototype generators operated in the garage attached to NIST's manufactured house, a test facility designed for conducting residential indoor air quality (IAQ) studies. This double-wide manufactured house is similar in size to homes commonly involved in fatal consumer incidents (Hnatov 2012). The garage tests described in the interim report documented reductions of 85 % to 98 % in CO concentrations due to emissions from two modified, prototype low CO-emission portable generators compared to an unmodified generator in this house for the scenarios tested.

Objective

The primary objective of this project was to measure and model CO emission and exposure, and O_2 depletion, resulting from portable generators operating in a home's attached garage under various use and environmental conditions in order to enable CSPC staff to analyze the safety implications of operating a portable generator under these conditions.

Contents of report

This report contains four main sections. The Experimental section describes the experimental methods and equipment, test house and shed, and generators used in the experimental portion of this project. The Shed Test Results section presents the results of testing three different generators operating in the test shed. The Garage Test Results presents the results of testing three different generators operating in the attached garage of the NIST test house. The Simulation section describes the CONTAM model of the test house, the model validation effort and the simulations used to extend the measurement results to other scenarios. The report also contains three appendices including one describing the uncertainty analysis of the measurements, one on instrument calibrations, and one with additional garage test results beyond those presented in the main body of the report.

Experimental

Methods and Equipment

Instrumentation

Gas concentrations were measured with two multi-gas engine exhaust analyzers (NOVA Analytics Model 7464), which are combination non-dispersive infrared (NDIR) and electrochemical sensor technologies, referred to as N1 and N2 in the rest of this report. These analyzers measured CO on two channels covering different ranges of 0 % to 1 % and 0 % to 10 %, CO_2 from 0 % to 20 %, hydrocarbons (as hexane) from 0 % to 2 % and O_2 from 0 % to 25 % with a reported accuracy of 1 % of full scale for all five channels. An electrochemical sensor CO analyzer (NOVA Analytics Model 7461, referred to as N3 in this report, measured CO over a range of 0 ppm_v to 2000 ppm_v and with a reported accuracy of 1 % of full scale. Two additional NDIR CO analyzers were used, a Thermoelectron Model 48 (referred to as TE) and a Rosemount Model 880A (referred to as RM), both with ranges of 0 ppm_v to 1000 ppm_v and reported accuracy of 1 % of full scale. Finally, a portable O_2 analyzer (Sybron Servomex O_2 Analyzer OA 580) was also used. Not all instruments were used during every test. Repeated calibrations during the test periods found that typical measurement uncertainties were consistent with the manufacturers' reported accuracies. See Appendix B for more detail on calibrations. To protect the analyzers from condensed water and/or soot particles, desiccant and high efficiency particulate air (HEPA) filters were used in the sampling system.

Air change rates were measured using tracer gas decay method (ASTM 2011). A pulse of sulfur hexafluoride (SF_6) was injected and allowed to mix before being measured with a gas chromatograph with an electron capture detector. A local weather station was used to measure ambient conditions. All gas concentration, air temperature, and humidity data were recorded by an automated data acquisition system. Nabinger and Persily (2008) provide more details on the SF_6, temperature, humidity, and ambient weather condition measurements including uncertainties.

Test House

The test house used in this study was a manufactured house located on the NIST campus, which was erected in 2002 (Nabinger and Persily 2008). An aerial view and floorplan of the house are shown in Figures 1 and 2. The house includes three bedrooms (MBR, BR2, and BR3), a living room (LR), a family room (FAM), a kitchen (KIT), and an attached garage. The house has a floor area of 140 m^2 and a volume of 340 m^3. The attached garage has a floor area of 36.5 m^2, a volume of 90 m^3 and was built as an addition to the house in 2007. The interior of the garage, including the ceiling, is finished with painted gypsum board. As part of the garage construction, the underlayment and siding of the exterior west wall of the house were removed and replaced with ¾ inch gypsum board on studs with fibrous glass batt insulation in the wall cavity.

Figure 1 Aerial view of NIST manufactured test house

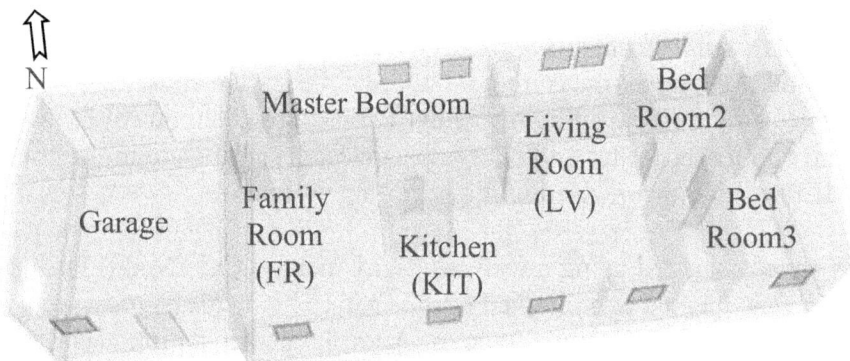
Figure 2 Floorplan of NIST manufactured test house

Measurements of gas concentrations were made at various points throughout the house using sample lines suspended 1.5 m above the floor in the center of each of the three bedrooms, the living room, the kitchen, and the family room, as well as five sample lines located near the four corners and center of the garage. The six individual living space locations were measured for one minute each in a repeating six minute cycle. The garage sample locations were measured separately, as well as a single mixed sample, the latter of which is reported here. Indoor air temperature and humidity were measured with sensors in each room of the house and on two opposite walls of the garage. The outdoor temperature and wind speed were measured at a weather station located about 6 m behind the house for tests conducted in 2008. For tests in 2010, wind speed data were collected from a weather station located on the roof of Building 226 on the NIST campus (about a mile from the test house), however, wind direction data were still collected at the local weather station.

Shed

Experiments were also conducted in a shed, with dimensions of 4.88 m (L) × 3.05 m (W) × 2.90 m (H), as shown in Figures 3 and 4, for the purpose of measuring the CO emission rate and O_2 consumption rate of the generators. The shed was used because it allowed better control of conditions than the garage and enabled a simplified single-zone analysis of the measurement results (see Shed Test Results section for description of analysis method). An explosion-proof

4

exhaust fan was installed in the wall opposite the door for shed ventilation at the end of a test and for quick exhaust during an emergency. This fan was also used in selected tests to create a high air change rate to obtain CO emission rates at near ambient conditions. The shed also had two operable windows at both sidewalls, which were adjusted to vary the air change rate.

Figure 3 Front view of the test shed

Figure 4 Inside view of the test shed

Separate sample lines were placed mid-height in the center of the shed (midway between the walls) for CO, O_2, and SF_6. Non-dispersive infrared and electrochemical sensor CO analyzers (N1, N2 and N3 described in the Instrumentation section) and a portable O_2 analyzer were used to measure CO and O_2 respectively. A gas divider/diluter was also used to dilute the sampled CO for the CO analyzer for some tests. The air temperature and humidity in the shed were measured at two locations near two sidewalls.

Generators and Loading

Generators were selected with electrical power output ratings in the size range most commonly involved in fatal consumer incidents, which is 5.0 kW to 6.5 kW (Hnatov 2012). Tests were conducted with three different generators that were configured in multiple ways. Two unmodified 'stock' (i.e., in their as-purchased condition) generators were tested. The first generator (referred to here as Gen B) has a full-load power rating of 5.5 kW with a 10 horsepower, carbureted, single cylinder gasoline engine and no specific CO emission control technology. This same generator was also tested by CPSC in a small chamber as reported on by Brown (2006).

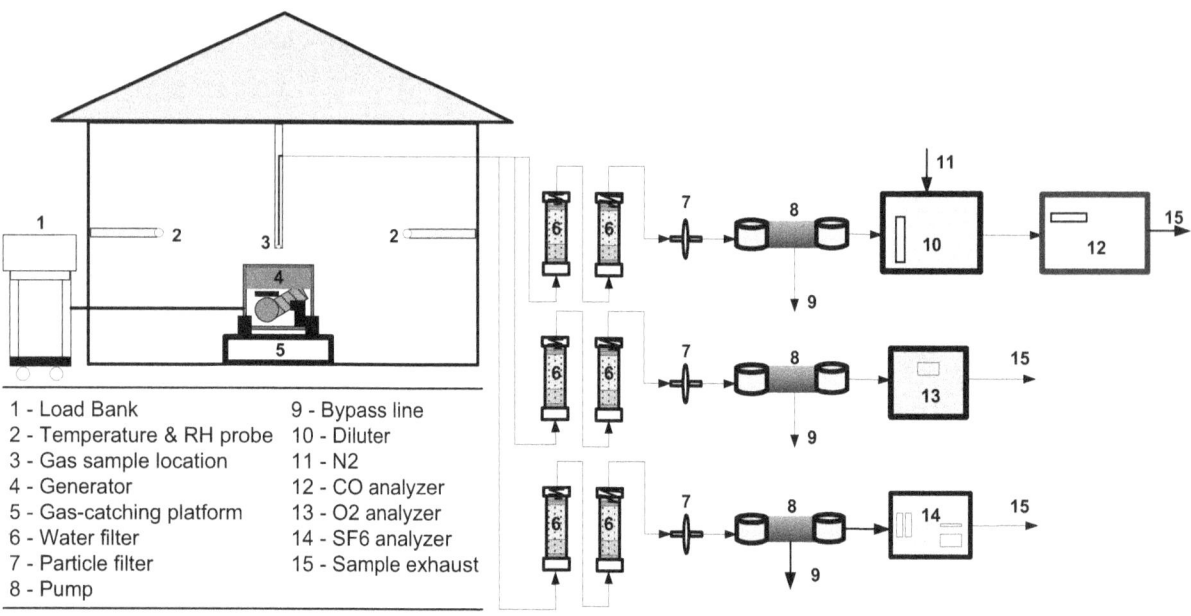

Figure 5 Schematic of experimental setup in shed

The second generator is powered by a carbureted 11 horsepower single-cylinder gasoline engine made by a different manufacturer than Gen B and has an advertised full-load electric power rating of 5.0 kW. This generator was tested in both unmodified condition (referred to as unmod Gen X) and as a modified low-CO emission prototype (referred to as mod Gen X). The unmodified generator operates at air-fuel ratios (AFR, ratio of mass of air to mass of fuel) in the range of 10 to 13 AFR depending on the load, which is common for small air-cooled carbureted engines. The modifications were made by the University of Alabama (UA) and included adding an engine management system (EMS) with associated sensors and actuators for electronic fuel injection (replacing the carburetor) and a muffler with a small catalyst integrated in it. The function of the EMS is to control ignition timing and fuel delivery through an engine control unit (ECU) microcomputer that receives input from a variety of system sensors. UA calibrated the ECU on the modified prototype to operate around a 14.6 AFR over the full range of loads. This AFR fuel control strategy is the primary means by which the prototype aims to achieve its reduction in CO emissions. The catalyst primarily targets reduction of oxides of nitrogen (NOx) and has relatively low catalytic activity because the EMS significantly reduces the available oxidation constituents in the exhaust stream.

For the third generator (referred to as Gen SO1), a model similar to Gen X was obtained which had the same model engine but with an alternator with an output rating of 7 kW. It was tested after UA modified it using the same fuel control strategy and largely the same emission control hardware that was used in mod Gen X. One difference is that Gen S01 had a different model ECU than that used on mod Gen X. Another difference noted during the testing is that its manufacturer included programming to maintain rich AFR operation until the oil temperature rose above approximately 60 $^{\circ}$C, resulting in an initial "spike" of CO when the engine was started cold. This ECU also includes an algorithm developed by UA that can be switched on or off by the test operator for testing purposes. The algorithm was intended to sense when the generator was operating in an enclosed space, based on engine operation parameters and when enabled, was intended to shut off the engine before a life-threatening CO hazard develops. All the tests with Gen SO1 that are reported in the main body of this report were performed with the algorithm disabled. Gen SO1 was also tested in a configuration with a muffler that did not contain a catalytic converter (referred to as the noncat muffler). The purpose of testing with the two different muffler versions was to measure the CO emissions produced in the engine due to the fuel control strategy alone (from tests with the noncat muffler) as well as get an indication of the catalyst's performance in further lowering those emissions (from the tests with the cat muffler). A full description of the prototype configuration of both mod Gen X and Gen SO1 is provided in greater detail in UA's report to CPSC (CPSC 2012).

To monitor prototype engine operation, generators mod GenX and Gen SO1 were outfitted with thermocouples and a Lambda sensor to measure AFR (ECM Model Lambda 5220, with an AFR range of 6 to 364 and reported accuracy of 0.2 for $12 <$ AFR < 18). The Lambda sensor and a thermocouple for measuring engine-out exhaust temperature were mounted through ports that UA provided on the exhaust manifold pipe between the engine and muffler. Cylinder head temperature was measured with a ring thermocouple mounted under the spark plug. Engine oil temperature was measured with a thermocouple inserted into the sump. For some of the tests, muffler and shroud temperatures were also measured, using thermocouples mounted directly on their surfaces at the hottest locations previously identified by UA with infrared cameras during their prototype tests.

The generators were operated using reformulated gasoline with 10 % ethanol obtained from the NIST motor pool, which is purchased to the same specification year-round. The generators were placed on a spill-catching platform in the middle of the garage (or shed) with the exhaust pipe pointing towards the garage wall adjoining the house.

A portable alternating current (AC) resistive load bank connected to the generator's 240-volt receptacle was used to draw electrical power and thereby act as a surrogate for consumer appliance loads. The load bank has manual switches in 250 W increments with a maximum setting of 10 kW. Some tests involved a constant load while others followed a cyclic load profile (see Table 1). This profile is similar to the load profile used by UA during the durability and emission testing of their low CO emission prototype generator. It was derived from a test cycle developed by industry to replicate typical in-use operation of small utility engines when used in all types of engine-driven products (CPSC 2012). Because the actual delivered power did not always match the load bank settings, particularly when oxygen depletion was occurring, the delivered power was measured and recorded during all tests.

Table 1 Hourly cyclic load profile

Load bank setting (W)	Duration (min)
no load	3
500	4
1500	18
3000	17.5
4500	12
5500	5.5

Shed Test Results

Few studies have been conducted directly on CO emission and O_2 consumption rates associated with gasoline-powered generators running indoors. Brown (2006) studied the CO emission rates from four different commercially-available generators in an enclosed experimental chamber, where air temperature and air change rate were controlled to provide different operating conditions. Also, the air change rates during these chamber tests were generally quite high compared with typical residences. Steady-state CO concentrations were found to be more than 7400 μL/L and O_2 levels as low as 18.5 %. CO emission and O_2 consumption rates at steady state were also calculated, and were found to be affected by O_2 level, generator loading, and/or chamber air temperature.

While generator tests conducted in laboratory chambers offer a high degree of control, they do not allow consideration of the real-world impacts of varying outdoor weather or the lower and variable air change rates that occur in residential structures. Alternatively, operating a generator in an enclosed space such as a garage or a storage shed, as opposed to a laboratory chamber, will be subject to uncontrolled temperatures and to lower ventilation rates determined by ambient weather conditions. To study generator CO emission rates under more realistic conditions including emission controls, tests were conducted in a single-zone space. This section reports on the measurement of CO emission rates from generators, operating in the unmodified carbureted configuration as well as in the low CO emission prototype configuration, (i.e. without and with CO emission controls) in a test shed located outdoors. The CO emission rates and O_2 consumption rates are used for the prediction of CO emission, migration and exposure in simulations.

Analysis Method

To model a generator running inside an enclosed single-zone space, a theoretical model was constructed based on Figure 6. This model was used to calculate CO emission and O_2 consumption rates from the concentrations measured during the tests. Assuming a gas component, C, is either generated ($S_C > 0$) or consumed ($S_C < 0$) in the zone, a differential mass balance equation for C during a period of $\Delta t = t_2 - t_1$ can be expressed as

$$\rho_{C,in} V_s \frac{dC}{dt} = S_C - \rho_{C,in} C Q_{out} + \rho_{C,out} C_{out} Q_{in} \qquad (1)$$

by assuming the following:
    the gas component, C, is an ideal gas,
    the concentration of C is uniform in the zone,
    ρ_C, S_C, and Q are constant during Δt, and
    the mass of fuel added from the generator to the zone air is neglected.
Note: See the Nomenclature list for all terms in the equations.

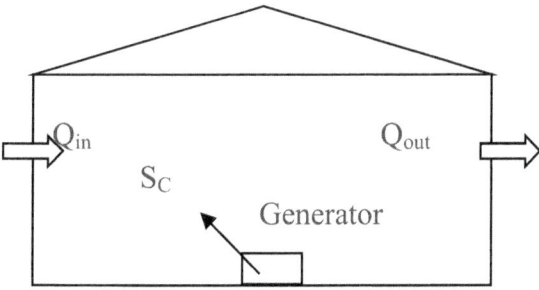

Figure 6 Schematic of a generator operating in a single-zone space

The density of C in Eq. (1) can be obtained by the ideal gas law. S_C is positive when referred to as S_{CO}, the emission rate of CO, and negative for S_{O2}, the consumption rate of O_2. In the experiments presented in this report, the concentration of a gas component, C, was measured by a gas analyzer, as described above. The air change rate, Q, was determined by using the tracer gas decay method using SF_6. Given that SF_6 may decompose at temperatures above 200 °C (Air Products 2006), the potential consumption of SF_6 in the generator engine, S_{SF6} (<0), was examined but was found to be insignificant (Wang and Emmerich 2010).

After determining the air change rate of the space from the decay rate of the SF_6 (see Wang and Emmerich 2010 for details), S_{CO} and S_{O2} can be solved from Eq. (1) for the time period of t_1 to t_2

$$S_{CO} = \rho_{CO,in} A_{out} V_s \frac{C_{CO,t2} - C_{CO,t1} e^{-A_{out}\Delta t}}{1 - e^{-A_{out}\Delta t}} \tag{2}$$

$$S_{O_2} = \rho_{O_2,in} A_{out} V_s \frac{C_{O_2,t2} - C_{O_2,t1} e^{-A_{out}\Delta t}}{1 - e^{-A_{out}\Delta t}} - \frac{\rho_{O_2,out} A_{out} V_s}{K_{m,\rho}} C_{O_2,out} \tag{3}$$

In addition, several gas concentration uniformity tests were conducted by collecting samples at five different locations in the shed. It was found that the thermal plume, which was driven by the heat from the running generator, mixed the shed very well throughout testing. The variations among the five sample locations were less than 5 % for SF_6, which met the uniformity requirement of 10 % in the ASTM standard (ASTM-E741-00 2006).

Generator B

Using the same generator (B), Brown (2006) found that CO emission rate was closely related to generator load and O_2 level in small chamber tests. In the current study, 13 tests of Gen B were conducted for load settings of 2.5 kW (half of the maximum load of the generator) and 5.0 kW (full load) for different air change rates (which result in different O_2 levels) and air temperature as shown in Table 2. Different air change rates were achieved by adjusting the shed windows to be opened with a 0.05 m × 0.2 m crack (OW) or completely closed (CW). Each test was also conducted at least twice to confirm the validity of the measurement data. Each test in Table 2 is named by its load setting, window adjustment and repetition, for example, 2.5kW-OW-1 indicates the first test running under a 2.5 kW load setting and with open windows. Note that Tests 6, 11, and 12 were conducted after the generator engine was serviced (S = serviced), which entailed changing the oil and cleaning the air filter. For Test 13, the shed was tightened to achieve a low air change rate (LA).

Table 2 Summary of Gen B Shed Test Results

Load setting (kW)	Test	Name	Run Time min	Max CO (μL/L)	Min O$_2$ (%)	CO Emission rates (g/h)	O$_2$ Consumption rates (g/h)	Air change rate (h^{-1})
2.5	1	2.5kW-OW-1	80	3190	19.3	340 to 900	3300 to 4800	6.5
	2	2.5kw-OW-2	63	5290	18.9	40 to 900	3300 to 4000	3.7
	3	2.5kw-CW-1	86	23950	16.9	1100 to 3100	4600 to 5900	2.6
	4	2.5kw-CW-2	101	23270	16.8	1000 to 3000	4200 to 5700	3.1
	5	2.5kw-CW-3	80	20090	17.1	880 to 2600	4100 to 5100	2.9
	6	2.5kw-CW-S1	106	8160	18.2	500 to 1400	4200 to 5100	3.6
5.0	7	5.0kw-OW-1	90	18210	17.9	120 to 3700	5700 to 7100	4.7
	8	5.0kw-OW-2	81	8420	19.1	1500 to 2600	4100 to 6500	7.6
	9	5.0kw-CW-1	62	21320	17.4	1100 to 3600	5600 to 6700	3.7
	10	5.0kw-CW-2	74	21200	17.1	1700 to 3000	4700 to 6400	2.9
	11	5.0kw-CW-S1	65	22470	16.9	1600 to 3500	5800 to 7600	3.0
	12	5.0kw-CW-S2	66	22280	17.0	70 to 3800	1400 to 7100	3.6
	13	5.0kw-CW-LA	86	25500	16.2	2000 to 3400	3900 to 6200	0.7

Table 2 shows large variations in CO and O$_2$ levels between tests with Gen B under the different constant load settings and window positions. The CO level reached only 3190 μL/L (note that μL/L are equivalent to the commonly used unit ppm$_v$) within 80 min in Test 1 when the generator was at half-load with the shed windows opened, but reached a maximum of 25500 μL/L for an 86-min run-time under full load in Test 13 when the shed was tightened. The shed O$_2$ concentration dropped significantly below ambient levels when operating the generator in this enclosed space. The largest decline occurred for Test 13, when the O$_2$ level was 16.2 % with the low air change rate and full loading. Table 2 also provides the measured air change rates and time-averaged weather conditions. When the shed was tightened in Test 13, the air change rate reached its lowest value, 0.7 h^{-1}, which explains it reaching the highest CO level and lowest O$_2$ level. For the rest of the tests, the shed window positions played an important role. The air change rates of open-window cases were all as high as or higher than those with closed windows. As a result, the opened window tests always had higher O$_2$ levels than those with closed windows for the same load. The CO levels in opened-window tests were generally lower than closed-window cases. Note also that the weather conditions during these tests varied significantly. The ambient temperature ranged from 4.0 °C to 17.2 °C and the wind speed from 1.1 m/s to 6.8 m/s. The variations of weather provided realistic generator operating conditions for these experiments as compared to controlled chamber experiments. Note that the CO and O$_2$

levels in Table 2 are only provided to indicate the range of test conditions, as these tests were primarily performed to determine CO emission and O_2 consumption rates for use in the simulation effort described later in this report.

As an illustration of individual shed tests, Figure 7 shows the CO and O_2 measured for Tests 1 and 13 from Table 2. The patterns of CO concentration in both tests are almost an inverse to that of the O_2 level for this unmodified generator. The CO level is low at the beginning of generator startup and increases steadily as the O_2 level drops. As the O_2 drops further and causes a very rich fuel mixture in the engine, CO reaches a maximum level. Test 13 in Figure 7 shows an extreme case in which the generator eventually produces a zero electrical load when the O_2 drops to around 16.4 %, although it was set at a full load and the engine crankshaft was still rotating.

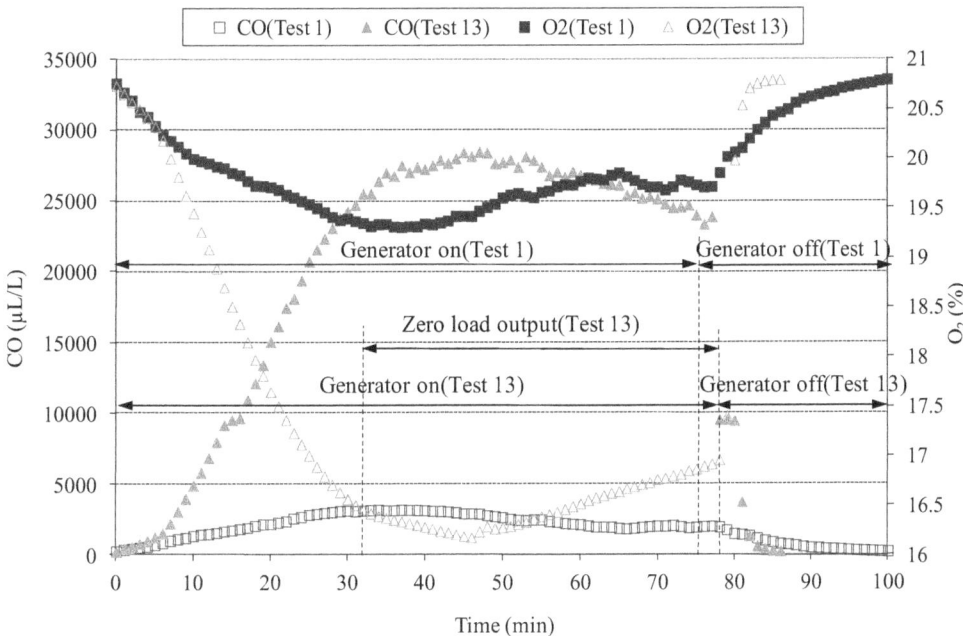

Figure 7 Measured CO and O_2 concentrations of Tests 1 and 13.

Figure 7 also shows that steady state was never reached for either test. A relatively stable period occurred at about 40 min for Test 1 and 45 min for Test 13, but they only held for a few minutes. These results differed from chamber experiments, where CO concentrations becomes constant after a period of time as complete steady state can be achieved under the controlled environment and higher air change rate. While these chambers tests are useful, the results from the shed confirm the importance of studying CO emission as a transient process under real weather conditions and more realistic air change rates to better understand generator performance in the field.

In order to generalize these test results to other conditions beyond this particular test facility, it is important to convert the results into CO emission and O_2 consumption rates. As seen in these tests, many factors can affect these rates directly or indirectly: space ventilation conditions, combustion conditions in the engine, O_2 level in the space, load setting, and the time over which the generator has been running.

Figures 8(a) and 8(b) illustrate how 5-min average CO emission rates and O_2 consumption rates (Δt = 5 min in Eqs. (7) and (8)) change with O_2 levels in the thirteen shed tests. Figure 8(a) shows that for both full and half load settings CO emission rates increase with decreasing O_2, reach maximum values when O_2 drop to about 17 % to 18 %, and then decline at lower O_2 levels. Under the extreme case of Test 13 (5.0kw-CW-LA), the CO rate decreases dramatically as the O_2 level reaches around 16.4 % with an electrical output of zero. Figure 8(b) shows an opposite trend of O_2 consumption rate compared to CO. The O_2 consumption rate decreases notably near 16.4 %, corresponding to the zero electrical output caused by O_2 deficiency. Given that the engine crankshaft was still rotating, the O_2 consumption rate was not quite zero, which explains the continual drop of O_2 in Test 13 of Figure 7, although the electrical output is zero.

The solid points in Figure 8 are data points for a half-load setting (2.5 kW) and the hollow ones for a full load setting (5.0 kW). As seen in previous small chamber test results in Brown (2006), a higher load setting generally results in more CO generated and O_2 consumed until the O_2 level reaches about 17 %, where data for full and half loads come together. This overlap corresponds to the drop in electrical output with the decrease of O_2. Note that Figure 8 also shows the calculated uncertainty for each data point of CO emission and O_2 consumption rates, which was mostly less than 20 % with a confidence level of 95 %. Appendix A discusses the uncertainty analysis in detail.

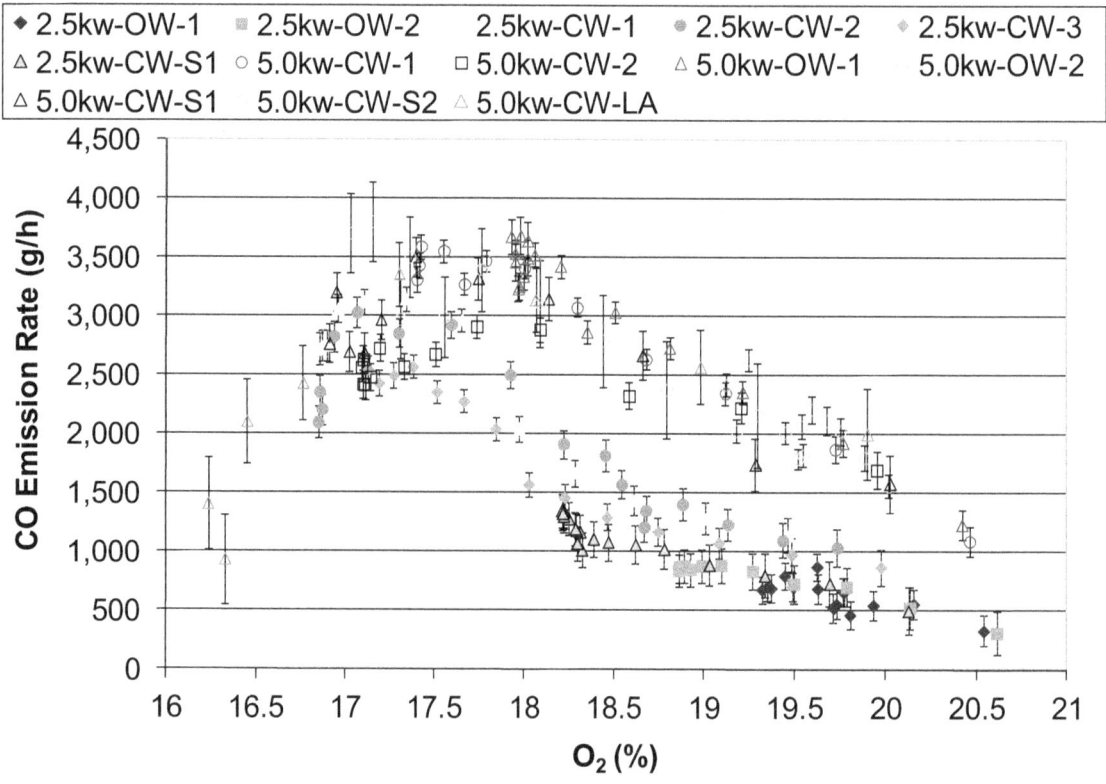

Figure 8(a)

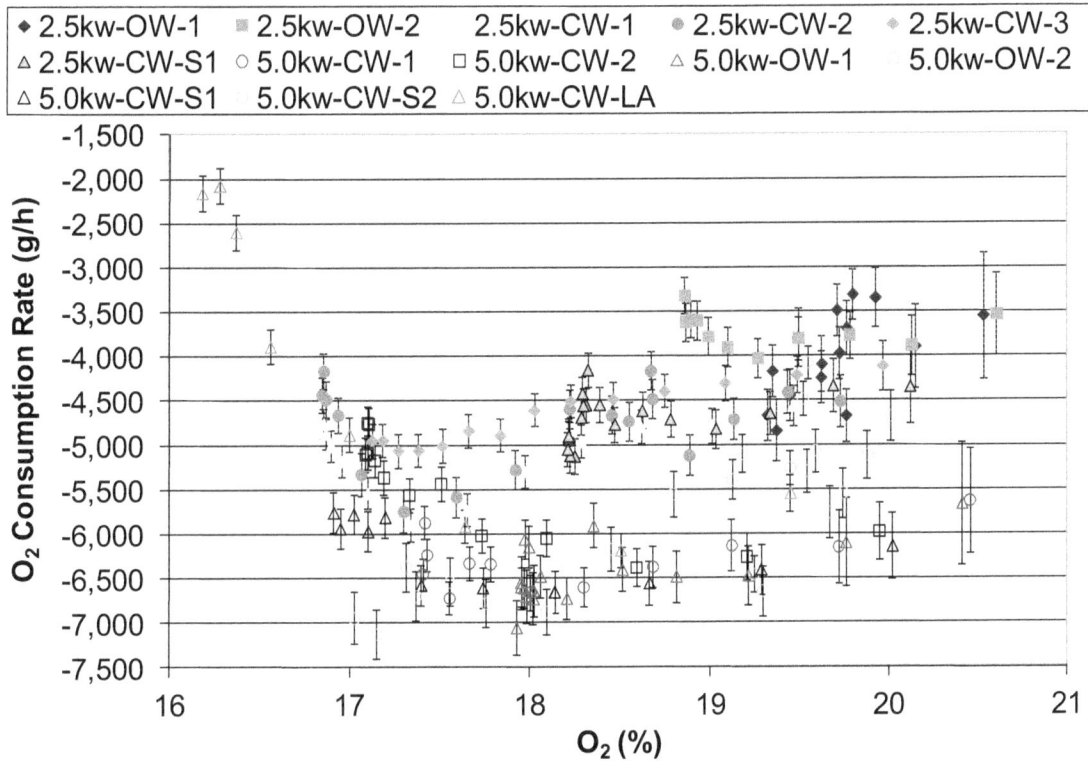

Figure 8(b)

Figure 8 Five-minute averaged (a) CO emission rates and (b) O_2 consumption rates at different O_2 levels for Gen B

Generator X

Generator X was tested in both unmodified and modified (low CO emission) prototype configurations. The primary difference between the tests with Gen X and Gen B was the generator loading. Gen X was tested at load points selected to approximately match the points of the load profile used by UA (See Table 1) during the durability and emission testing of their low CO emission prototype generator.

Figures 9(a) and 9(b) present the CO emission rates and O_2 consumption rates as a function of O_2 levels for unmodified Generator X. Although the tests of Gen X and Gen B were not identical, Figure 9(a) show similar results in that the CO emission rates range from a low of around 500 g/h at near ambient conditions to a high of nearly 4000 g/h as O_2 approaches 17 %. Unlike Gen B, however, the emission rate is only clearly load-dependent when the O_2 drops below about 19 %. Fewer tests were performed on Generator X below 17 % O_2 but the results indicate a similar drop off in CO emissions due to poor engine performance under these conditions.

13

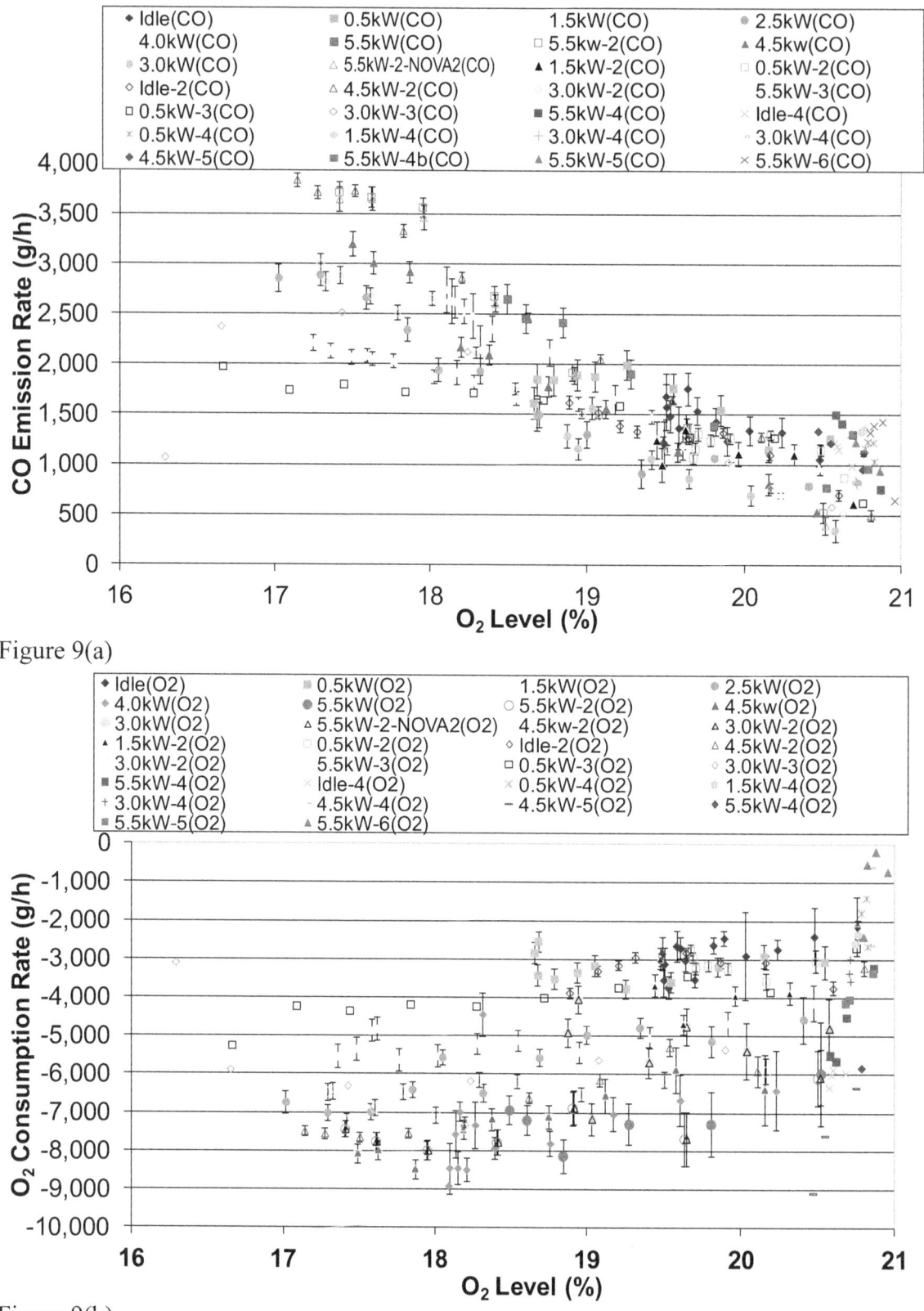

Figure 9(a)

Figure 9(b)

Figure 9 Five-minute averaged (a) CO emission rates and (b) O_2 consumption rates at different O_2 levels for unmodified Generator X

Figures 10(a) and 10(b) present the CO emission rates and O_2 consumption rates as a function of O_2 levels for modified Generator X. Although modified Gen X was not tested as many times as unmodified Gen X, comparing Figure 10(a) to Figure 9(a) shows the dramatic reduction in CO emission rates due to the low CO emission modifications included on the prototype. Most of the modified Gen X CO emission rates were well below 500 g/h. Although not enough low O_2 tests were performed to be conclusive, the CO emission rates at the highest loads did tend to increase as O_2 dropped.

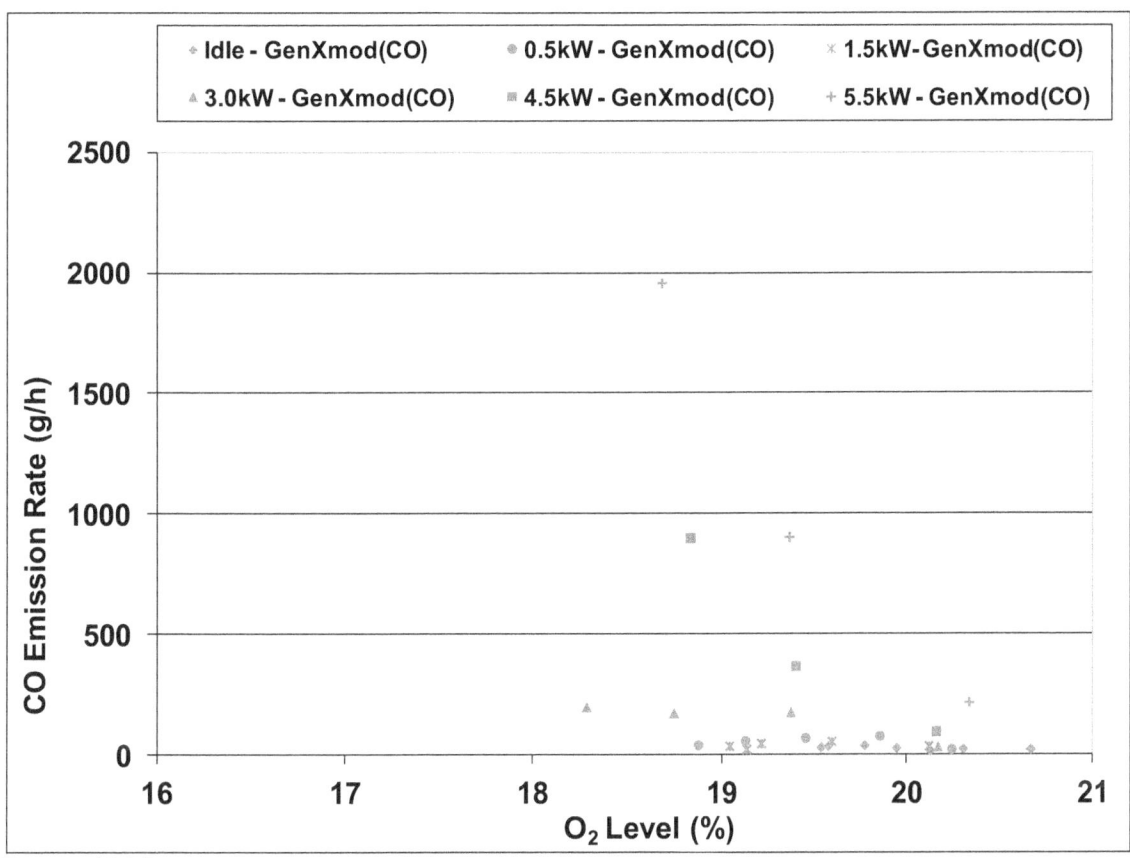

Figure 10(a)

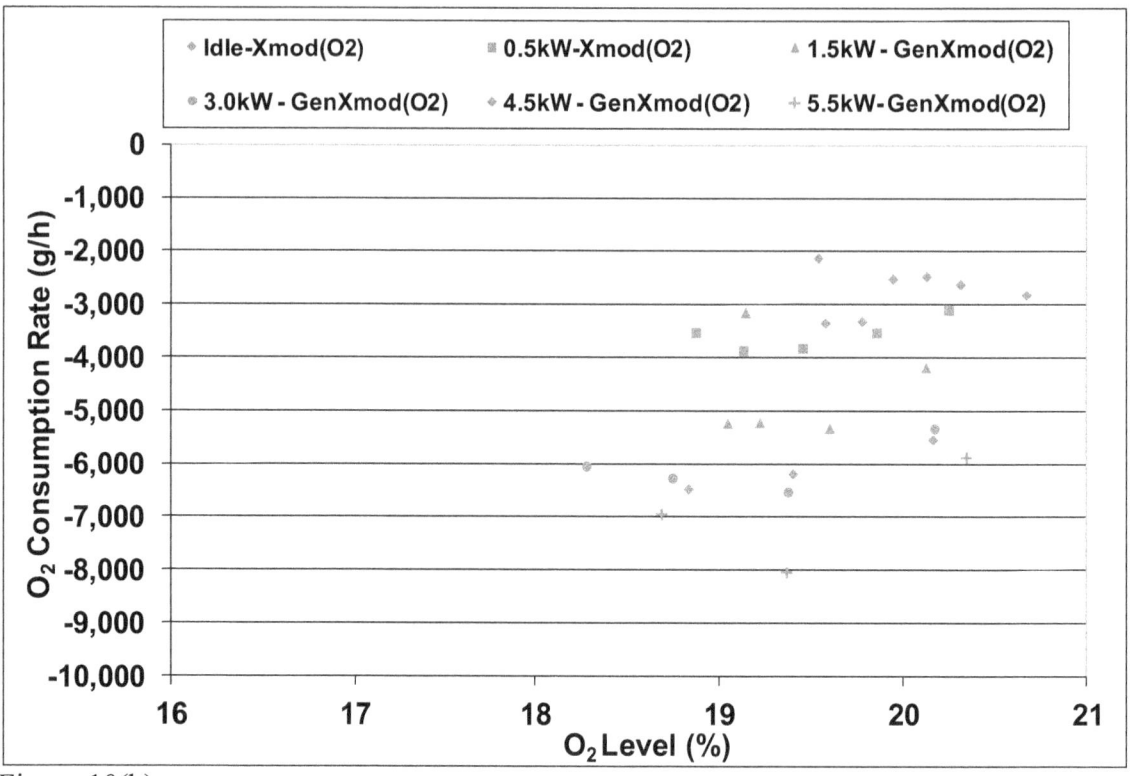

Figure 10(b)

Figure 10 Five-minute averaged (a) CO emission rates and (b) O_2 consumption rates at different O_2 levels for modified Generator X

Generator SO1

Generator SO1was tested at the same load points as Generator X but was not tested in an unmodified configuration. Instead, it was tested in two modified configurations – with and without a catalyst integrated in the muffler (referred to as cat muffler and noncat muffler, respectively). Figure 11(a) and 11(b) present the CO emission rates and O_2 consumption rates as a function of O_2 levels for Generator SO1 with the cat muffler (referred to as Gen SO1 cat). Comparing Figure 11(a) to Figure 10(a) shows that Gen SO1 cat performed somewhat better than modified Gen X. All measured CO emission rates for Gen SO1 cat were well below 500 g/h, and no trend toward higher emission rates was seen as O_2 levels dropped to 18 %. However, as with Gen X, no tests were performed at levels as low as 17 %.

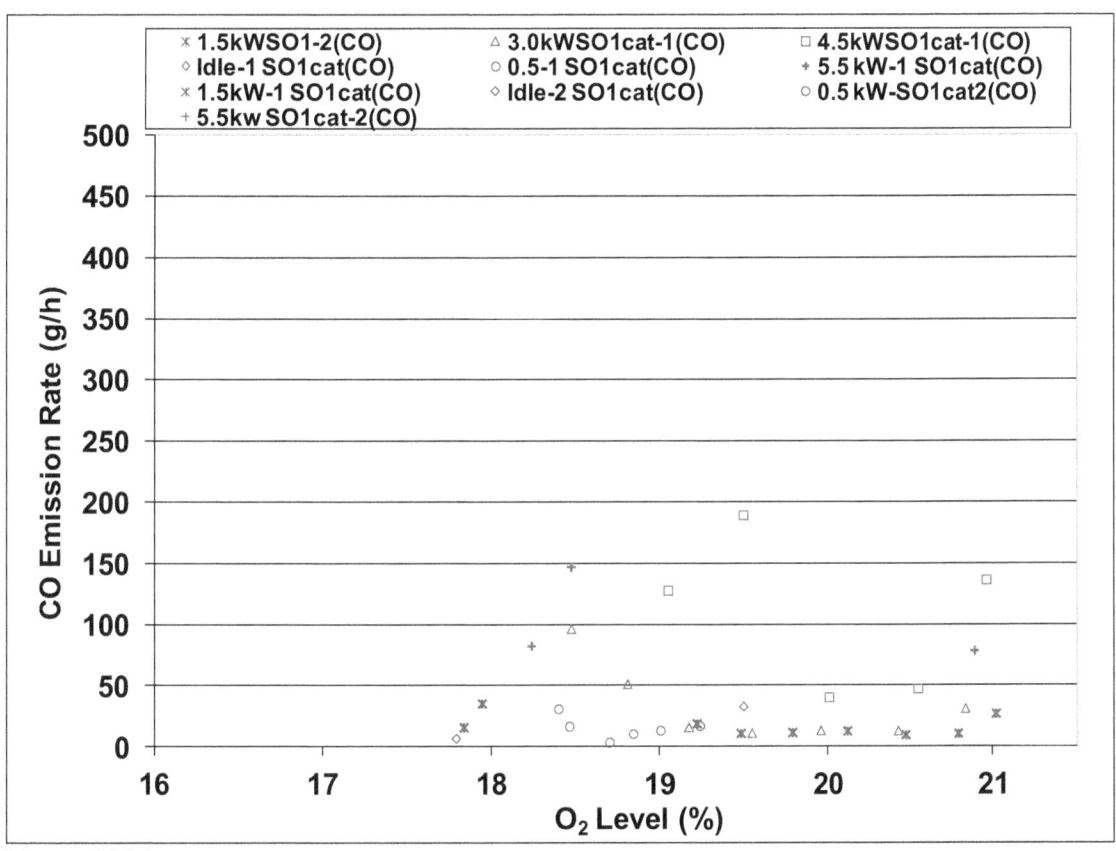

Figure 11(a)

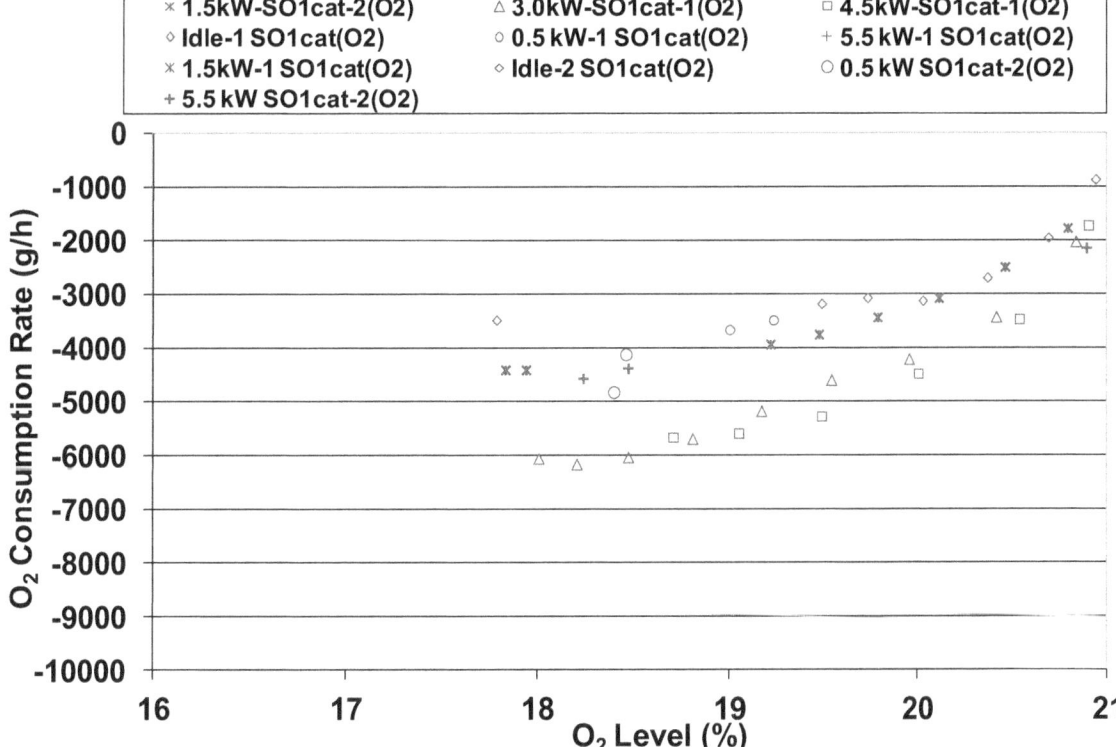

Figure 11(b)

Figure 11 Five-minute averaged (a) CO emission rates and (b) O_2 consumption rates at different O_2 levels for Generator SO1 with cat muffler

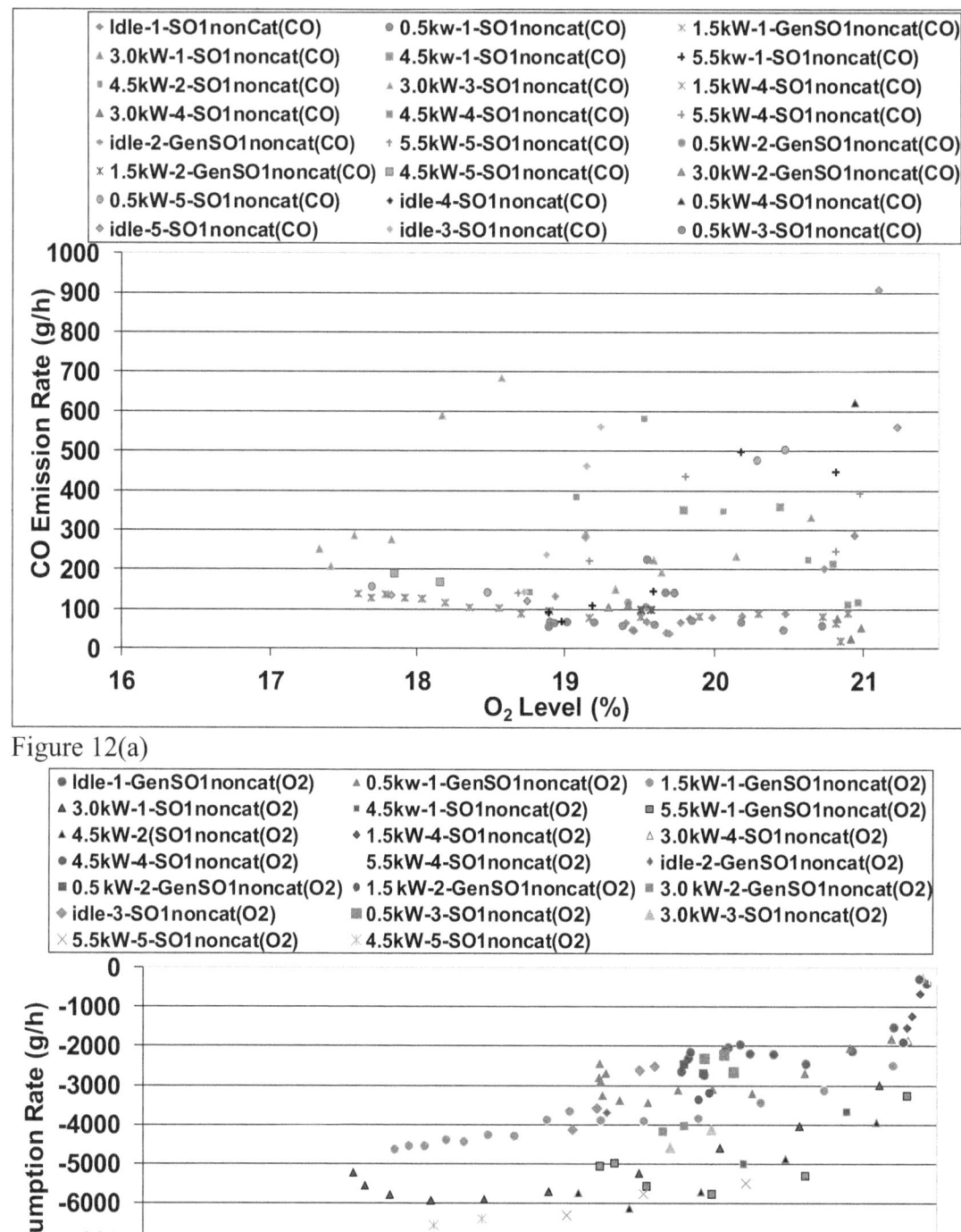

Figure 12(a)

Figure 12(b)

Figure 12 Five-minute averaged (a) CO emission rates and (b) O_2 consumption rates at different O_2 levels for Generator SO1 with noncat muffler

Figures 12(a) and 12(b) present the CO emission rates and O_2 consumption rates as a function of O_2 levels for Generator SO1 with the noncat muffler (referred to as Gen SO1 noncat). Comparing Figure 12(a) to Figure 11(a) shows that Gen SO1 noncat had higher CO emission rates than Gen SO1 cat. However, the measured CO emission rates for Gen SO1 noncat were still substantially lower than the emission rates of the unmodified generators, and no trend toward higher emission rates was seen as O_2 levels dropped close to 17 %.

Discussion of Shed Tests

A series of experiments were conducted to measure CO emission and O_2 consumption rates of portable, spark-ignited, gasoline-powered electric generators (in both unmodified carbureted configuration and prototype low CO emission configuration) operated in a single-zone shed. These tests were used to derive the CO emission and O_2 consumption rates to be used as inputs to a model validation effort, as well as for the simulation analyses conducted to examine the potential performance of the low CO-emission prototype under a wider range of operating conditions. For two different unmodified generators (i.e., without CO emission controls), it was found that CO emission ranged from a low of around 500 g/h at near ambient O_2 levels, to a high of nearly 4000 g/h as O_2 approached 17 %. The rates of CO generation and O_2 consumption were affected by multiple parameters, with the O_2 level and the actual electrical output of the generator being two of the most important. Tests performed below 17 % O_2 showed a drop off in CO emissions due to poor engine performance under these conditions. Tests of two modified prototype low CO emission generators (i.e., with CO emission controls) showed reductions of CO emissions of over 90 % depending on the specific emission controls and operating conditions.

Garage Test Results

NIST also conducted a series of tests to provide empirical data to characterize the emission and transport of CO due to operating portable gasoline-powered generators in an attached garage. This section of this report presents data from these tests of both unmodified and UA-modified prototype generators operated in the garage attached to NIST's manufactured test house. Note that these results apply to the specific units tested and that other units, modifications, houses and test conditions may produce different results.

Testing Configurations

Testing was conducted under seven different test house configurations to evaluate their impacts on the buildup of CO in the garage and its transport into the different rooms in the house. These configurations included two different garage bay door positions (fully closed or open nominally 0.6 m), two connecting door settings between the garage and the family room (fully closed or open nominally 5 cm), and two house central heating, ventilating, and air conditioning (HVAC) fan settings (on or off). All internal house doors were kept open throughout all tests.

There were multiple purposes in conducting tests under these different configurations. The garage-house door positions directly affect the rate of engine exhaust transfer from the garage into the house. The status of the HVAC fan, which circulates the interior air throughout the rooms of the house, affects the CO distribution within the house. The fan operation also affects the house air change rate due to air distribution duct leakage within the crawl space (Nabinger and Persily 2008). It is also relevant to consider the fan status, even when there is a power outage, because the consumer may use the generator to provide power to the home's central heating system, which includes providing power to the HVAC fan. Another reason for testing under different garage door and garage-house door positions is that, with the generator operating in the garage, it is possible that the engine will consume the oxygen in the garage at a faster rate than the rate at which natural air change replenishes oxygen. The degree of either door's opening will impact whether or not the garage's oxygen level can be maintained at ambient level and, if not, how low it will drop. Testing with different door positions enabled observations of the effects of different oxygen levels on generator engine performance. Variations in house configuration can be found in CPSC's investigation reports of fatal CO poisonings involving generators (Hnatov 2010). These reports include cases in which consumers were aware of the CO poisoning hazard but attempted to provide what they considered "proper ventilation" by operating the generator in a partially-open garage. A bay door opening of 0.6 m was selected in part based on it being within the range of openings that can be modeled using the multizone airflow and IAQ model CONTAM. The house door opening of 5 cm was selected because it is a reasonable opening to allow the passage of an extension cord from the generator into the house.

Table 3 includes a summary of the tests conducted including information on the generator tested, the test house configuration (defined by door positions and fan status), a test identification code, the date the test was conducted, the average ambient temperature and wind speed, and the CO analyzers used. As described in the Instrumentation Section, several different analyzers were used during the tests to span the full range of CO concentrations, but data is presented only from the analyzer considered most appropriate for the CO concentration range in each test. All tests listed in Table 3 were "cold start" tests. Additional garage tests are reported in Appendix C.

Table 3 Tests Conducted in Attached Garage

Generator	House Configuration	Garge bay door	Garage to house door	HVAC fan	Test ID	Date	Outdoor Temp (°C)	Wind speed (m/s)	CO analyzers in garage	CO analyzers in house
unmod GenX	1	Closed	Open	OFF	B	04/22/08	20.1	6.5	N1	N2, N3
modGenX	1	Closed	Open	OFF	O	04/02/10	22.0	6.5	N2, N3	N1, R1
SO1	1	Closed	Open	OFF	N	04/01/10	19.9	6.3	N2, N3	N1, R1
unmod GenX	2	Open	Closed	OFF	F	05/06/08	22.8	7.7	N1	N2, N3
modGenX	2	Open	Closed	OFF	R	04/12/10	19.9	6.7	N2, N3	N1, R1
SO1	2	Open	Closed	OFF	T	04/14/10	13.4	6.9	N2, N3	N1, R1
unmod GenX	3	Closed	Open	ON	I	05/15/08	22.8	7.4	N1	N2, N3
SO1 with noncat muffler	3	Closed	Open	ON	Z	05/05/10	28.3	6.7	N2, N3	N1, R1
unmod GenX	4	Closed	Closed	ON	J	05/21/08	18.2	9.6	N1	N2, N3
SO1	4	Closed	Closed	ON	W	04/29/10	17.8	9.5	N2, N3	N1, R1
unmod GenX	5	Closed	Closed	OFF	D	04/30/08	12.2	8.2	N1	N2, N3
SO1 with noncat muffler	5	Closed	Closed	OFF	AH	05/13/10	15.6	6.5	N2, N3	N1, R1
unmod GenX	6	Open	Open	ON	G	05/07/08	25.1	7.0	N1	N2, N3
SO1	6	Open	Open	ON	U	04/22/10	20.4	7.8	N2, N3	N1, R1
unmod GenX	7	Open	Open	OFF	K	05/23/08	13.84	7.0	N1, T1	N2, N3
SO1 with noncat muffler	7	Open	Open	OFF	V	04/23/10	15.8	6.5	N2, N3	N1, R1

Results

Figures 13 through 28 show the measurement results for all 16 tests listed in Table 3, including CO concentration in the house and garage, O_2 concentration in the garage, and the measured electric load supplied by the generator. In all tests, the generator was started at time 0 and was manually shut off by the test operator using a wireless switch that interrupted the engine's ignition. The data in the figures are plotted up until the time mechanical venting was initiated, which typically immediately followed generator shut-off. In some tests, where time and circumstances permitted, natural decay was allowed to occur after the generator was stopped, before mechanical venting was initiated. In those tests, the natural decay is plotted.

Figures 13a, 13b, and 13c show the results for Test B, which was a three hour test of unmod Gen X in Configuration 1 (garage bay door closed, garage access door to house open nominally 5 cm, and the house central HVAC fan off). Since it was a three hour test, the hourly cyclic load profile in Table 1 was repeated three times. At the end of the third cycle, the generator was stopped, and the garage was mechanically vented.

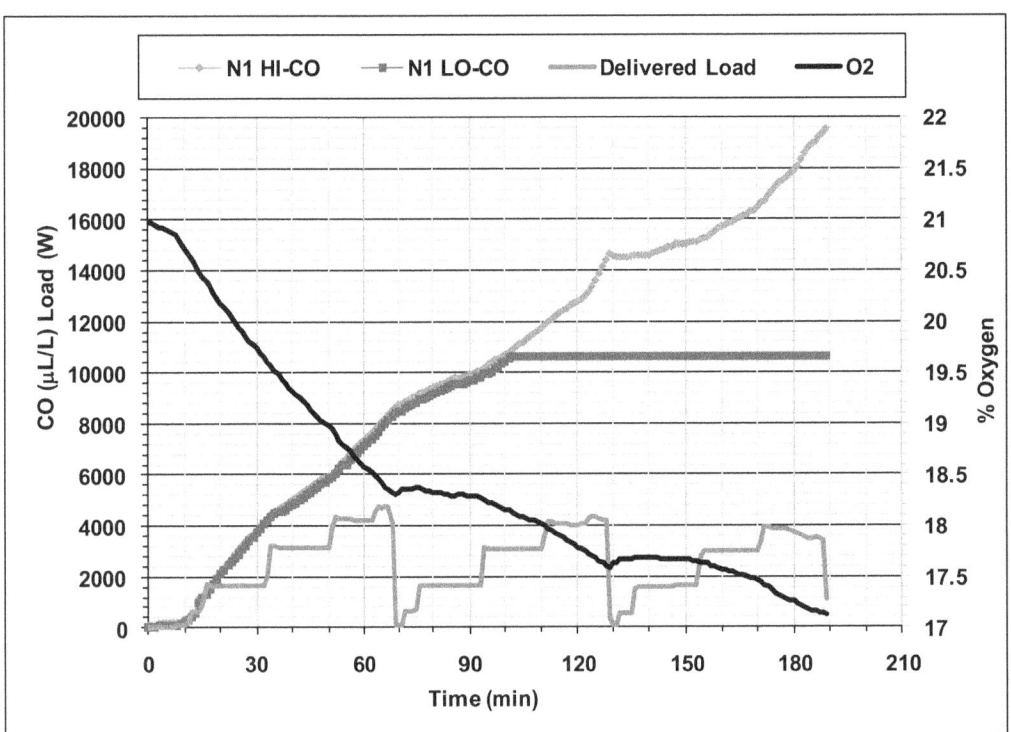

Figure 13a CO and O_2 concentrations in the garage and measured load for Test B (unmod Gen X, Configuration 1)

Figure 13a shows the concentration of CO in the garage reached a peak of over 19,500 µL/L and the volume fraction of O_2 in the garage dropped by 3.8 % to nearly 17 % when the generator was stopped. It also shows that in the first load cycle, the delivered electrical output was less than the load bank settings for the two highest loads in the load cycle, 4500 W and 5500 W, which were applied when the oxygen was already below 19 %. As the oxygen continued to drop in the subsequent load cycles, the delivered power for these load points decreased further.

22

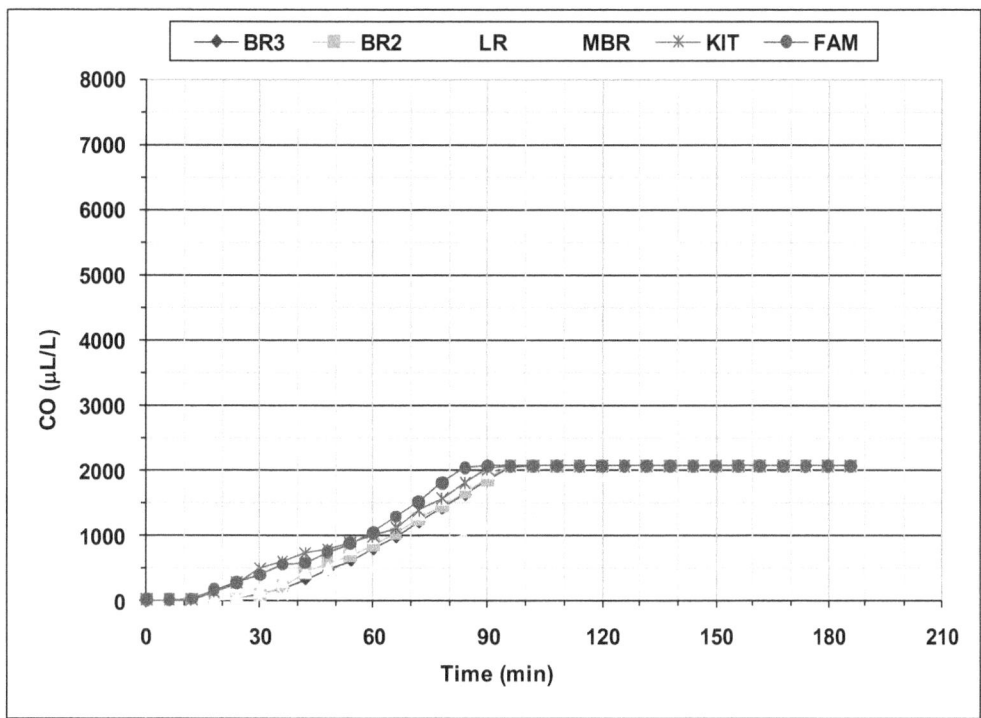

Figure 13b CO (ppm range) concentrations in the house for Test B (unmod Gen X, Configuration 1)

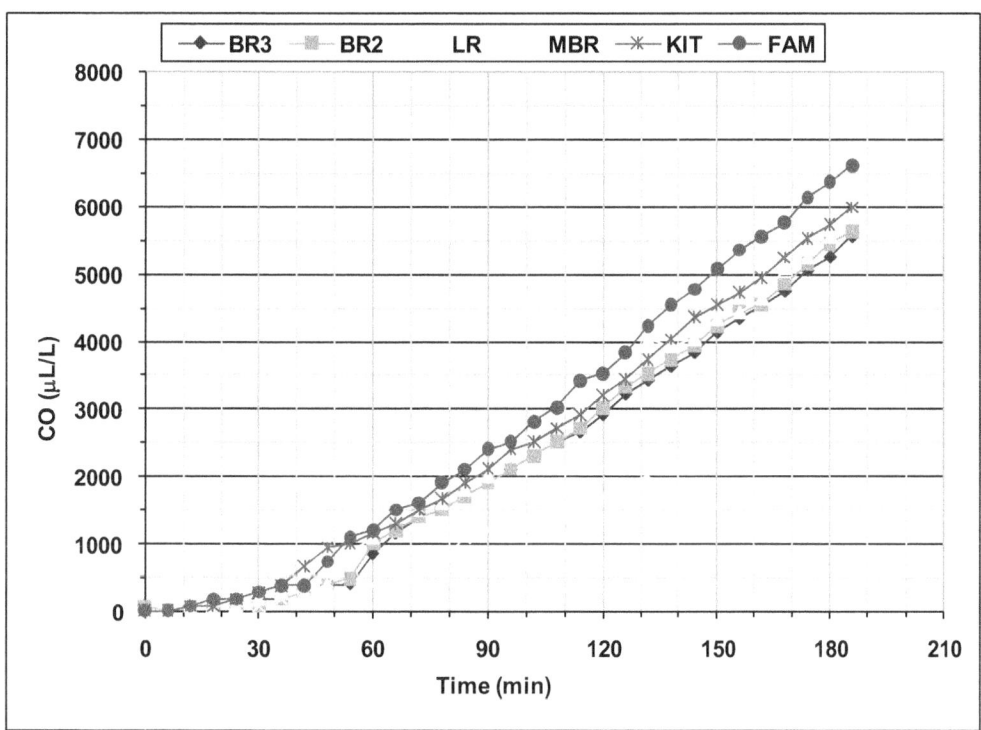

Figure 13c CO (high range) concentrations in the house for Test B (unmod Gen X, Configuration 1)

Figures 13b and 13c show the CO concentration in six rooms of the test house (see Figure 2 for room locations) as measured with the N3 'ppm range' (where the CO concentration plot plateaus at the instrument's 2000 μL/L limit) and N2 'high range' CO instruments, respectively. The CO

23

reached a peak concentration of over 6500 μL/L in the family room, with peak concentrations in the other rooms ranging from about 3500 μL/L to 6000 μL/L. As described in the Instrumentation section, samples were taken for one minute at each location on a rotating six minute cycle.

Figures 14a, 14b, and 14c show the results for Test O, which was a four and a half hour test of mod Gen X with the same test house configuration (Configuraiton 1) as used in Test B of unmod Gen X (shown in Figure 13). After the generator was stopped, the garage and house were mechanically ventilated.

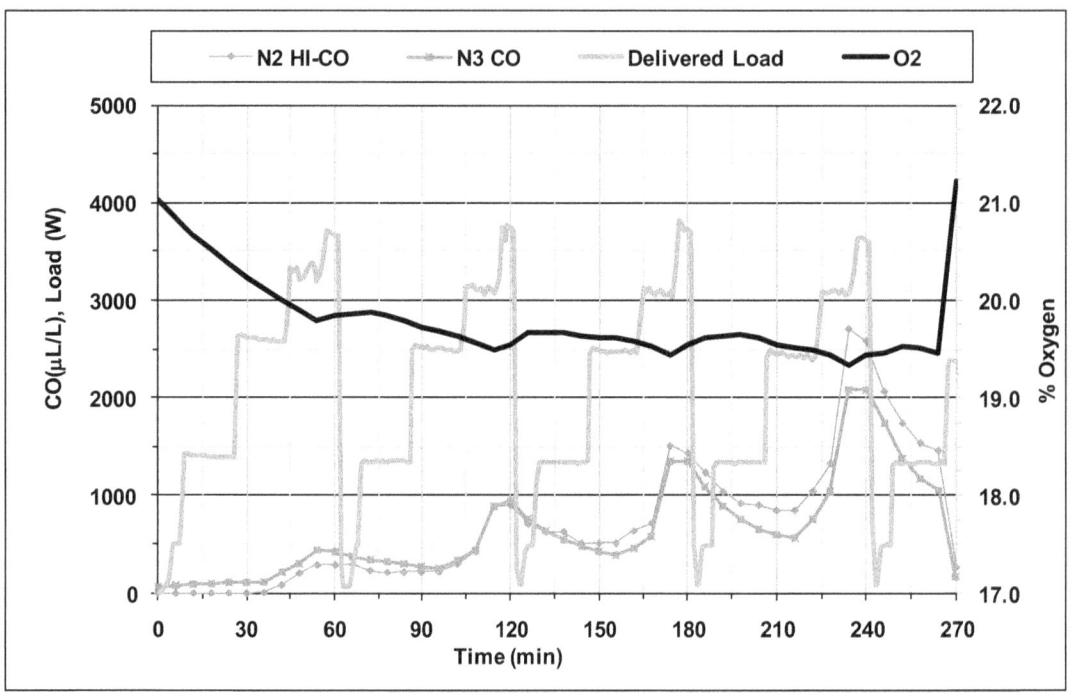

Figure 14a CO and O_2 concentrations in the garage and measured load for Test O (mod Gen X, Configuration 1)

As shown in Figure 14a, the garage CO concentration reached a peak of nearly 3000 μL/L while the garage O_2 concentration dropped by 1.7 % to 19.5 % after completing the fourth cycle of the load profile. Note that the ppm instrument (N3) briefly topped out at this time, about 230 min into the test. Also, the initial O_2 concentration is shown as slightly above 20.9 % for this and some other tests due to the limitations in instrument accuracy. The generator was intentionally stopped midway through the fifth load cycle.

At three hours into this test, the garage CO concentration was approximately 1400 μL/L. Under fairly similar ambient conditions between this test and Test B, this CO concentration is a 93 % reduction compared to that measured with unmod Gen X in Test B. In that case, the garage CO was over 19,500 μL/L at the same time during the test.

In the first load cycle, as the oxygen dropped, the delivered electrical output was less than the load bank settings for the three highest loads in the load cycle, 3000 W, 4500 W, and 5500 W. While the electrical output stayed near constant for the four cycles, the CO levels increased progressively and the oxygen decreased slightly with each additional cycle.

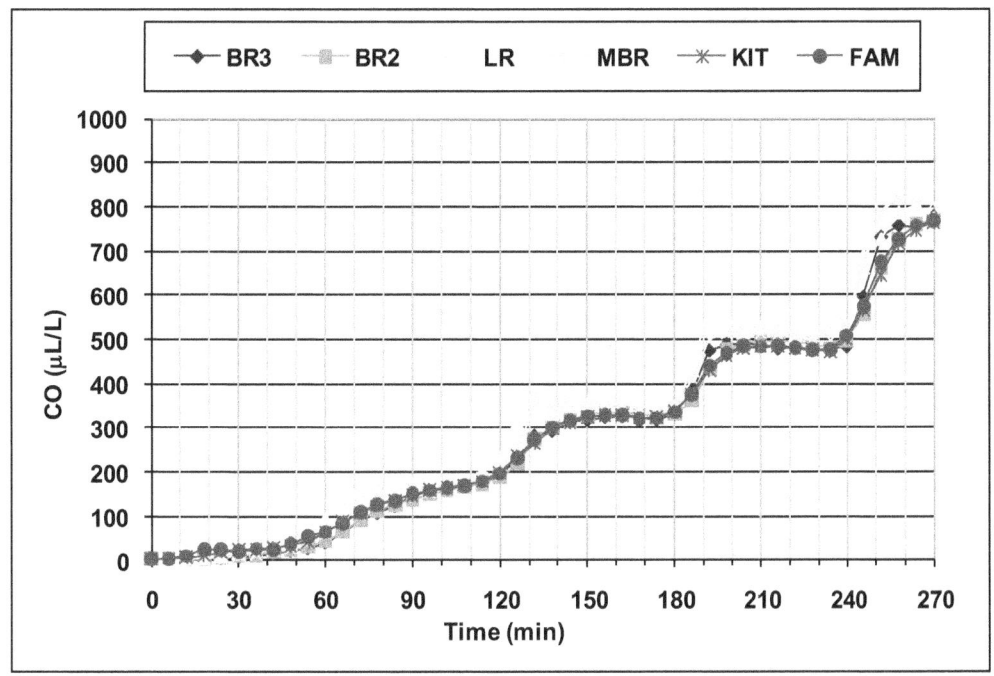

Figure 14b CO concentrations in the house for Test O (mod Gen X, Configuration 1)

As seen in Figure 14b, the peak CO concentration throughout the house was about 800 µL/L, with a relatively uniform distribution in all the rooms despite the HVAC fan being off. By comparison, unmod Gen X in Test B produced a peak concentration of over 6500 µL/L in the family room.

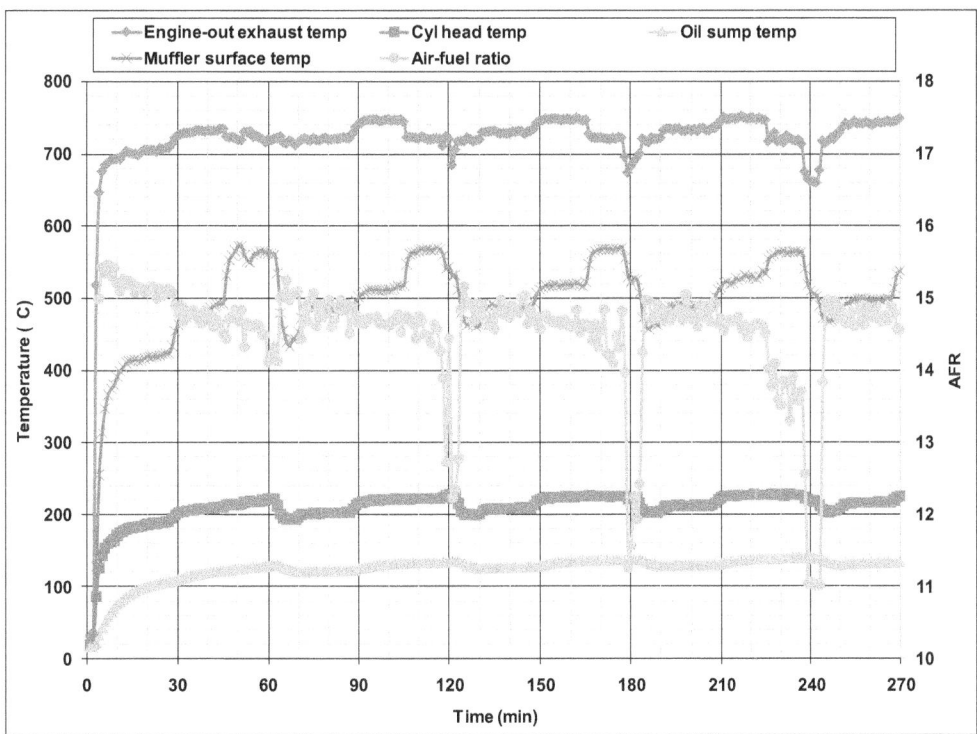

Figure 14c Temperatures and AFR measured in Test O (mod Gen X, Configuration 1)

The AFR (provided as a general indicator of the prototype engine's performance for this and other tests) and temperatures measured on modGen X during Test O are shown in Figure 14c. During this test, the engine performed off design with AFR largely ranging from around 14 to around 15.4 during each load cycle and reducing to rich operation when transitioning between the load cycles as well as during the high loads.

Figures 15a, 15b, and 15c show the results for Test N, which was a two hour test of Gen SO1 with the same test house configuration used in Test B of unmod Gen X and Test O of mod GenX (Configuration 1). This test was terminated earlier than planned after a fuse blew on the load bank after 114 min of operation, dropping half the load. The generator was turned off 138 min after it was started. A natural decay period of 45 min was included after the generator was stopped, followed by mechanical venting.

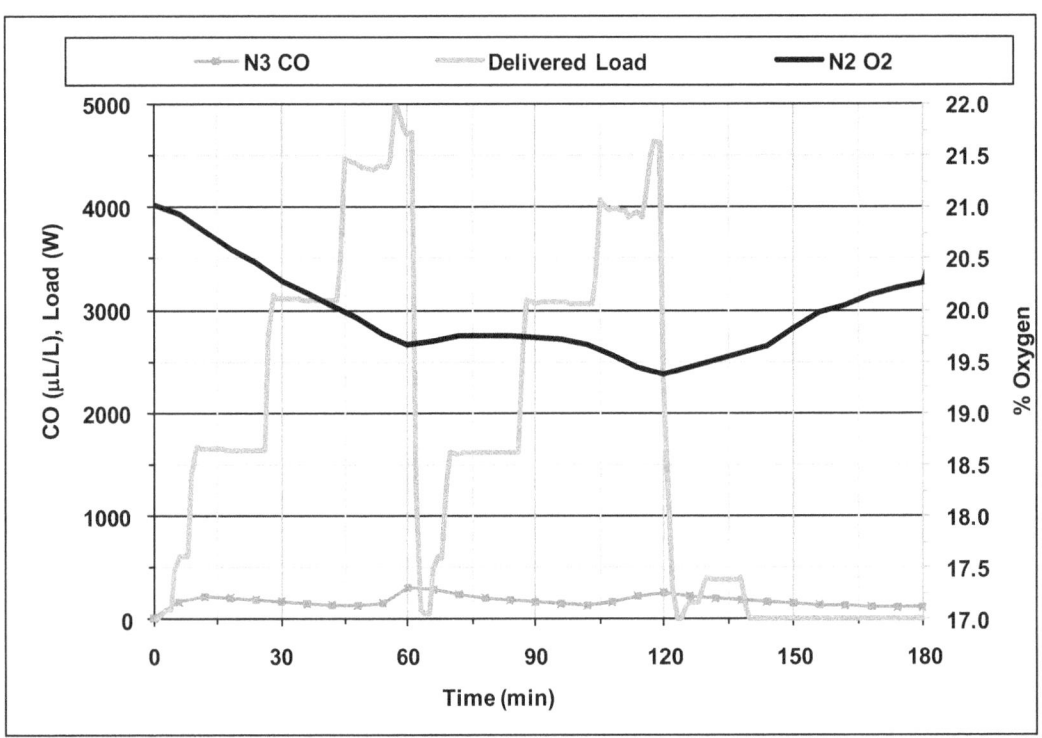

Figure 15a CO and O_2 concentrations in the garage and measured load for Test N (Gen SO1, Configuration 1)

As shown in Figure 15a, there was an initial increase of CO to almost 220 μL/L in the first 12 min after the generator was started. This rise is due to the rich operation upon cold engine start until the oil warms and the ECU transitions to the calibrated AFR fuel control. This initial increase is observed at the start of each of the tests with Gen SO1. The garage CO concentration reached a peak of around 300 μL/L, and the garage O_2 concentration dropped by 1.6 % to 19.4 % before the generator was stopped. The garage CO concentration after two hours is about 98 % lower than the concentration at two hours with unmod Gen X in Test B, which was about 13,000 μL/L. In the first load cycle, as the oxygen dropped, the delivered electrical output was less than the load bank setting for the highest load in the load cycle, 5500 W. This difference increased in the subsequent load cycle as the oxygen level decreased. Comparing the performance of mod Gen X (Figure 14a) and Gen SO1 (Figure 15a) shows that, under similar conditions (Configuration 1), Gen SO1 resulted in significantly lower CO concentrations at the 2 h mark.

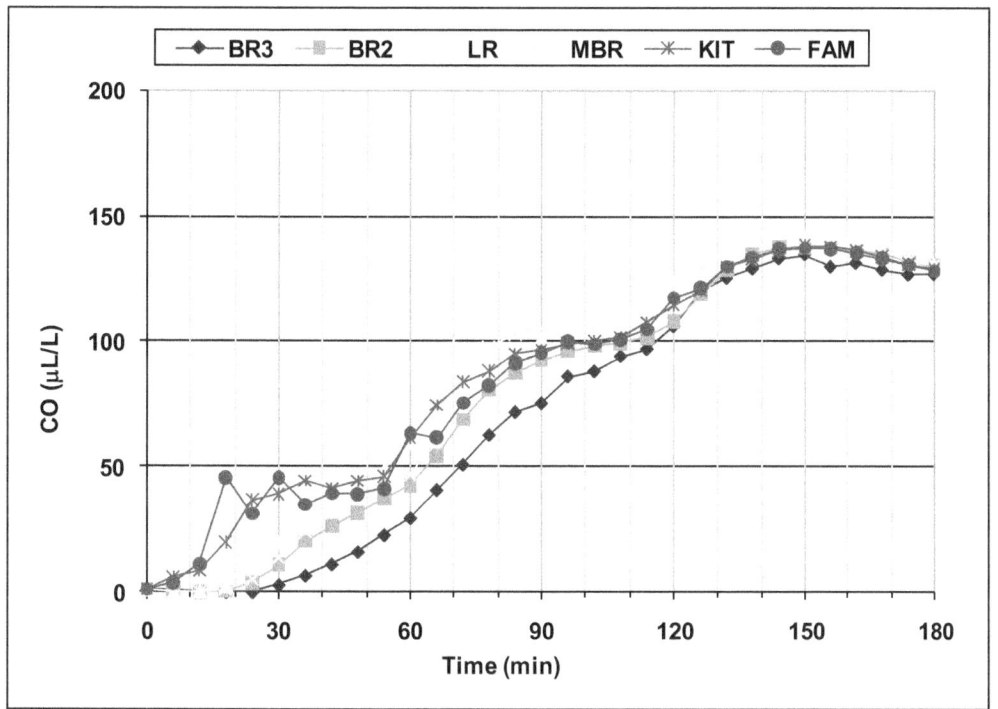

Figure 15b CO concentrations in the house for Test N (Gen SO1, Configuration 1)

As shown in Figure 15b, the concentration throughout the house was about 130 μL/L when the generator was stopped after 114 min. There is a relatively even CO distribution among the rooms in spite of the HVAC fan being off. For the following 45 min in which the exhaust was allowed to naturally decay, CO continued to infiltrate from the garage into to the house, slightly increasing the house concentration to about 140 μL/L before the concentration began dropping. By comparison, unmod Gen X in Test B produced a peak concentration of over 3500 μL/L in the family room after 2 h.

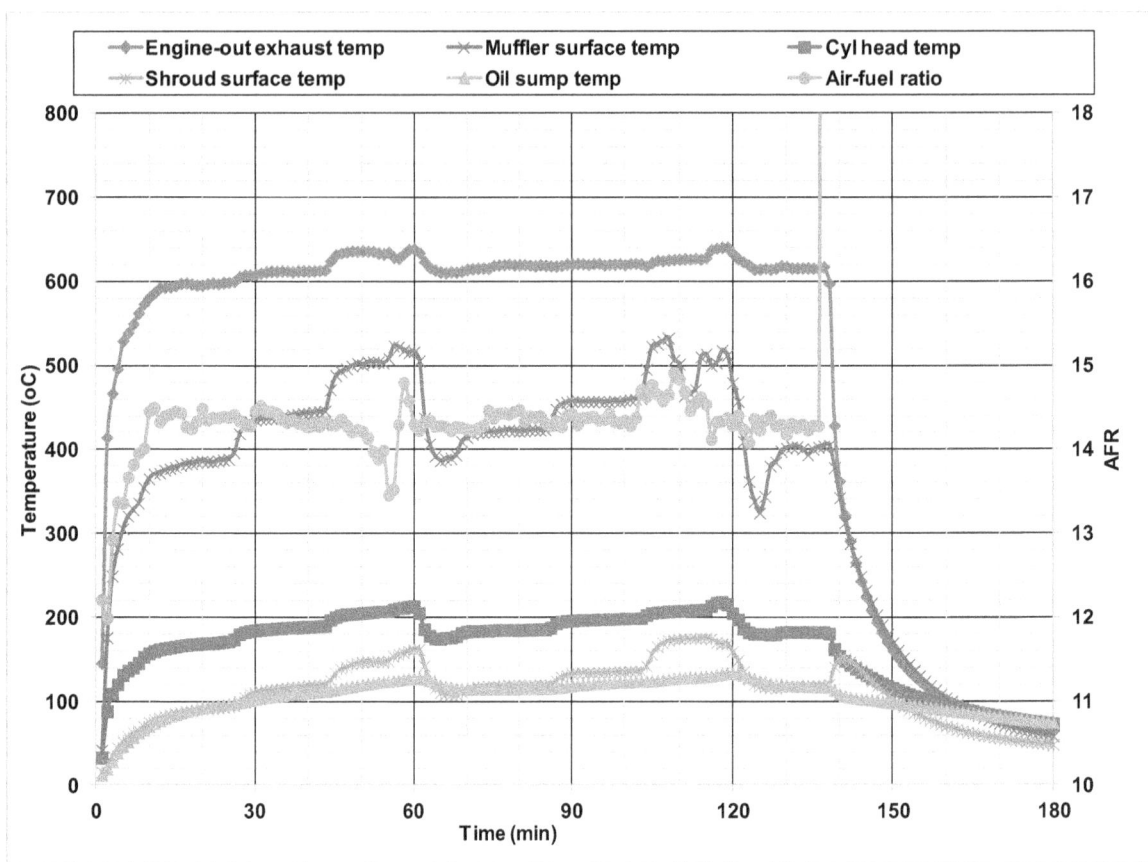

Figure 15c Temperatures and AFR measured on Gen SO1 in Test N (Gen SO1, Configuration 1)

The AFR and temperatures measured on Gen SO1 during Test N are shown in Figure 15c. With the exception of two periods of AFR excursion after the engine warmed up (i.e., after approximately 10 min), the engine operated at the calibrated AFR of 14.5 as the oxygen level dropped. The spike in AFR at the end of the test occurred when the engine was turned off.

Figures 16a and 16b show the results for Test F, which was a four hour test of unmod Gen X with Configuration 2 (garage bay door open, garage access door to house closed, and house central HVAC fan off). After the generator was stopped, the garage concentration was allowed to naturally decay for one hour before the garage and house were mechanically vented.

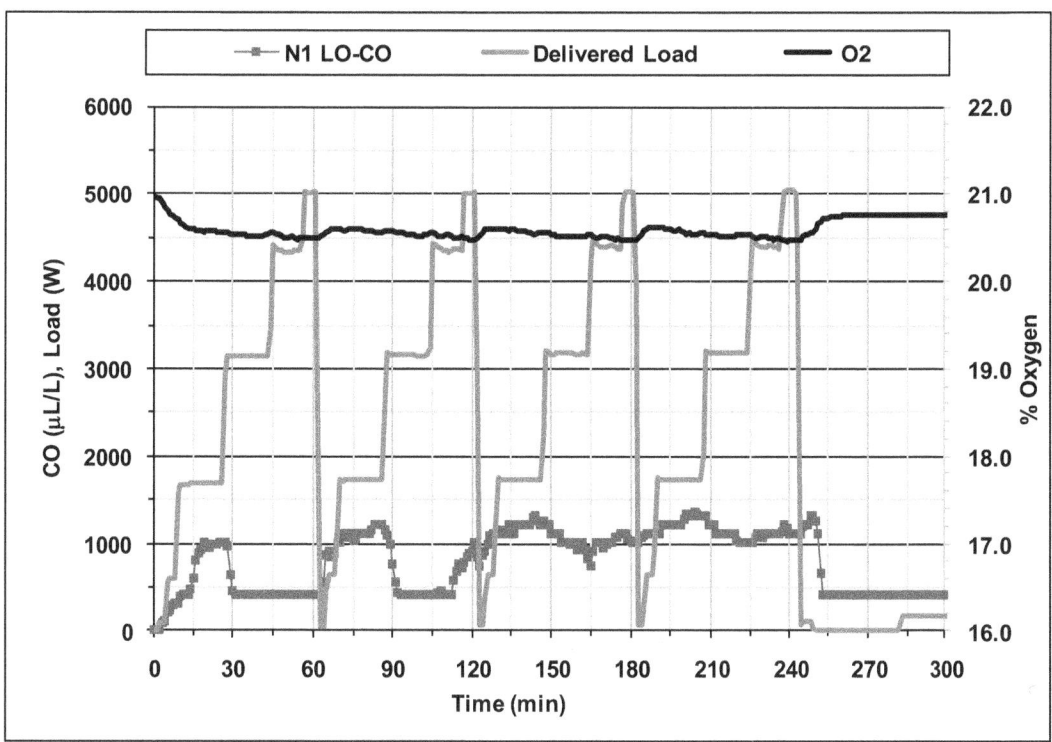

Figure 16a CO and O_2 concentrations in the garage and measured load for Test F (unmod Gen X, Configuration 2)

The garage CO concentration peaked during each load cycle during the 1500 W load bank setting. The peak concentration rose slightly in each load cycle, reaching a maximum concentration somewhat under 1500 μL/L in the fourth load cycle. For this test, the garage was not instrumented with a low concentration CO analyzer, and the instrument uncertainty is large relative to measured concentrations below 500 μL/L.

During this test, with the garage bay door open, the garage oxygen level decreased only slightly, down by 0.5 % to 20.5 %, and the delivered electrical output was consistent during each cycle, largely meeting the load bank setting with the exception of the 5500 W setting.

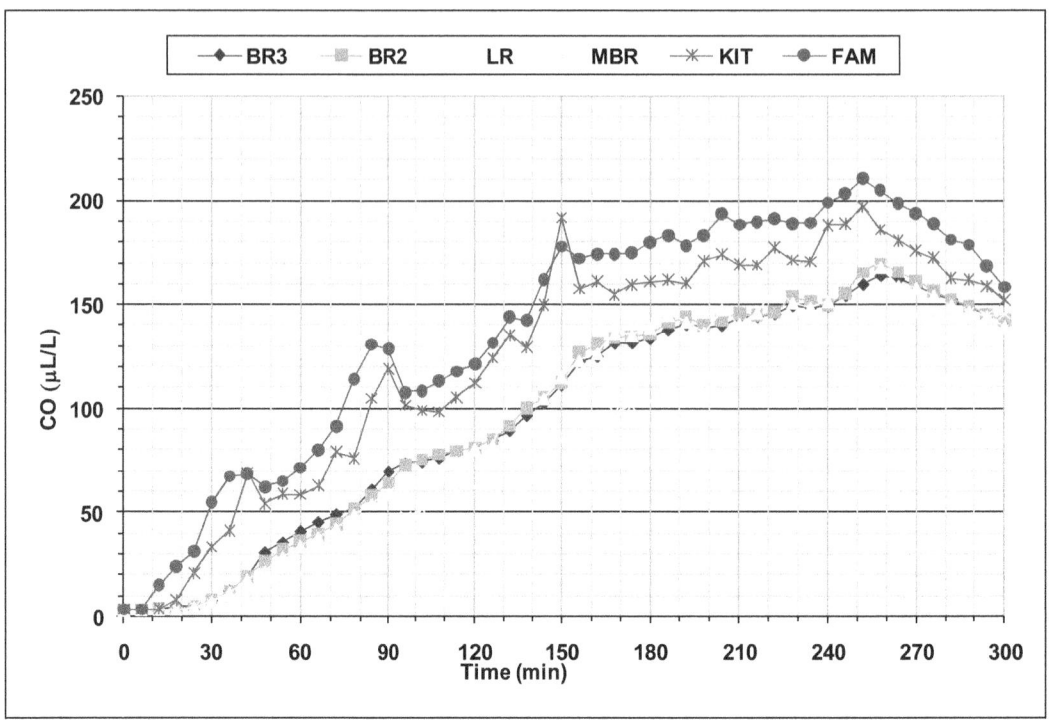

Figure 16b CO concentrations in the house for Test F (unmod Gen X, Configuration 2)

As shown in Figure 16b, the maximum house CO concentration was measured in the family room at just over 200 µL/L about 15 min after the generator was stopped after a 4 h runtime. The master bedroom had the lowest peak concentration among all the rooms, reaching just over 150 µL/L about 30 min after the generator was stopped.

Figures 17a, 17b, and 17c show the results for Test R, which was a four hour test of mod Gen X with the same test house configuration as used in Test F of unmod Gen X (Configuration 2). Mechanical venting was initiated right after the generator was stopped.

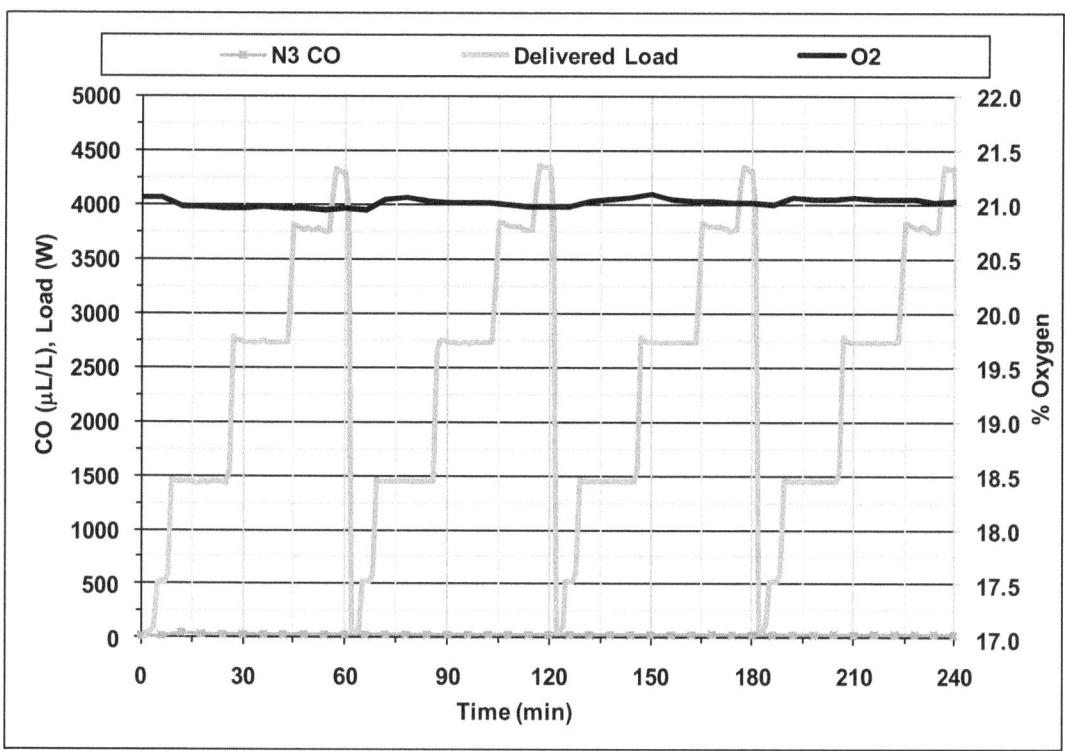

Figure 17a CO and O_2 concentrations in the garage and measured load for Test R (mod Gen X, Configuration 2)

As seen in Figure 17a, the garage CO concentration was nominally steady at 30 μL/L (though the uncertainty of the instrument is large relative to this level), and the oxygen stayed nominally at ambient throughout the test. This is about a 98 % reduction in CO compared to the nearly 1500 μL/L measured with unmod Gen X in Test F.

The delivered electrical output was less than the load bank settings for the three highest loads in the load cycle, which occurred with no significant oxygen depletion. After this test, the unit was thoroughly inspected, including the wiring between the generator head and the 240-volt receptacle, but no anomalies were found. During UA's development of this prototype, they observed on several occasions that these wires and associated connector melted.

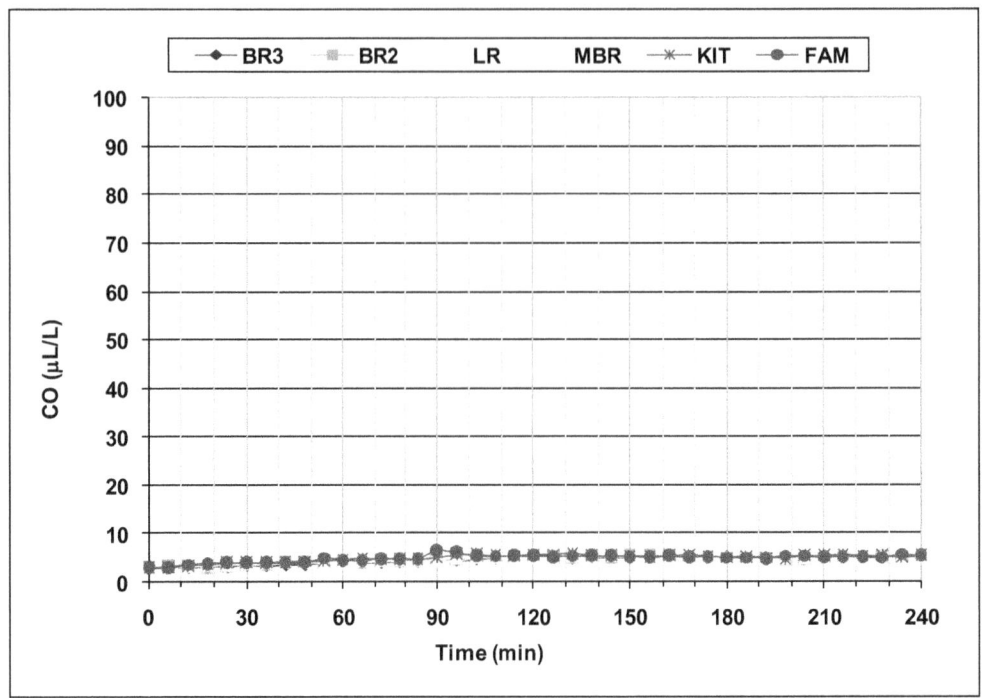

Figure 17b CO concentrations in the house for Test R (mod Gen X, Configuration 2)

The CO concentration throughout the house was nominally steady at 5 μL/L in all rooms throughout the test (though the instrument uncertainty is large relative to this concentration). By comparison, unmod Gen X in Test F produced a maximum CO concentration in the family room at just over 200 μL/L, corresponding to a reduction of around 98 % in Test R.

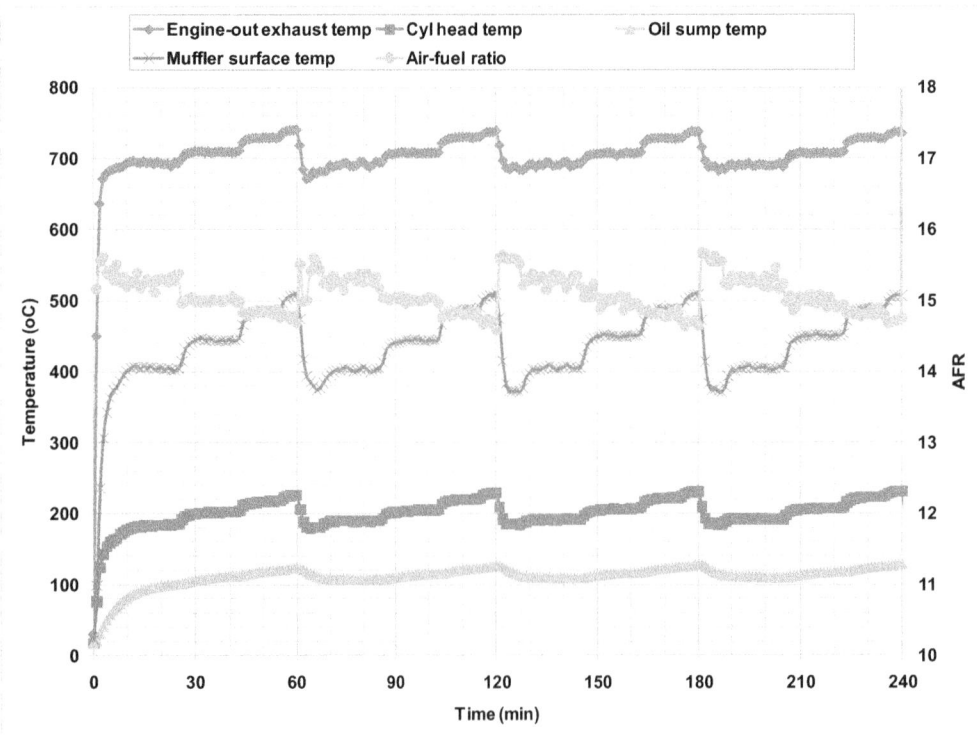

Figure 17c Temperatures and AFR measured in Test R (mod Gen X, Configuration 2)

The AFR and temperatures measured on modGen X during Test R are shown in Figure 17c. During each load cycle, the engine primarily ran lean, with the AFR ranging from about 14.5 to 15.6. The spike in AFR at the end of the test corresponds to when the engine was turned off.

Figures 18a, 18b, and 18c show the results for Test T, which was a three hour test of Gen SO1 with the same test house configuration as used in Test F and Test R of unmod Gen X and modGenX respectively (Configuration 2). The generator was stopped when a circuit breaker on the 240-volt receptacle tripped. Mechanical venting was initiated right after the generator was stopped.

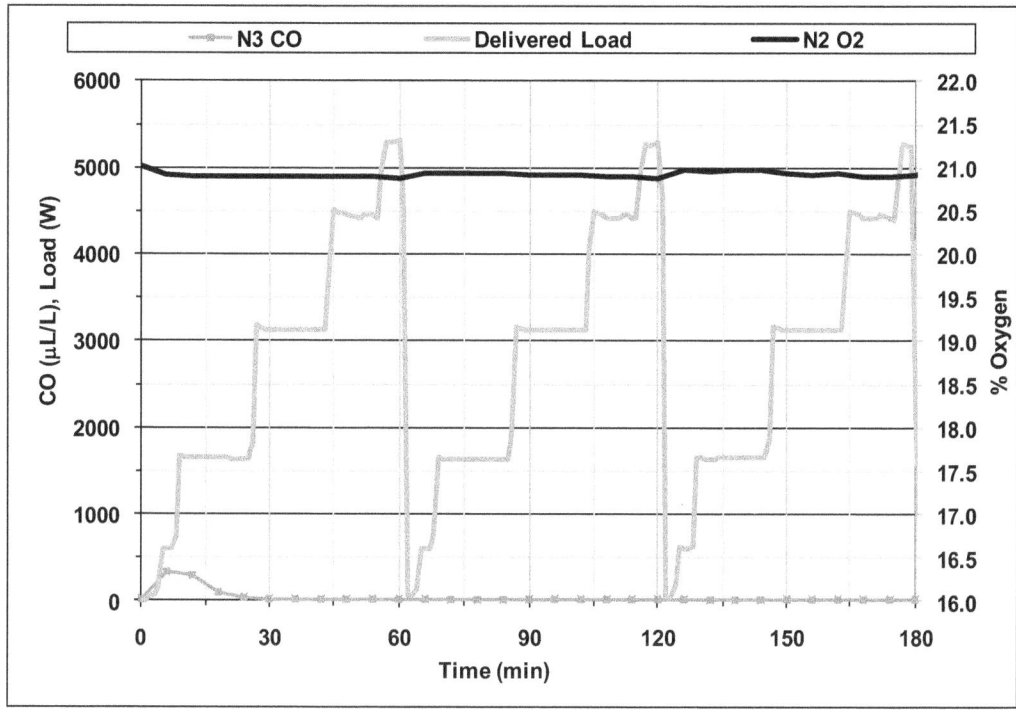

Figure 18a CO and O_2 concentrations in the garage and measured load for Test T (Gen SO1, Configuration 2)

As shown in Figure 18a, there was an initial spike of CO in the garage of over 300 µL/L when the engine was started and as the oil warmed before operation transitioned to the calibrated AFR. The CO concentration then dropped and maintained a level of about 20 µL/L throughout the test (though the uncertainty of the instrument is large relative to this level). With the garage bay door open, the garage oxygen level stayed nominally at ambient. With the exception of the early peak, this CO concentration is over a 98 % reduction compared to the peak garage CO measured with unmod Gen X in Test F. Throughout the test, the delivered electrical output was consistent during each cycle, largely meeting the load bank setting with the exception of the 5500 W setting. Comparing the performance of mod Gen X (Figure 17a) and Gen SO1 (Figure 18a) shows that, for Configuration 2, both generators resulted in similar low CO concentrations after the initial spike in Test T.

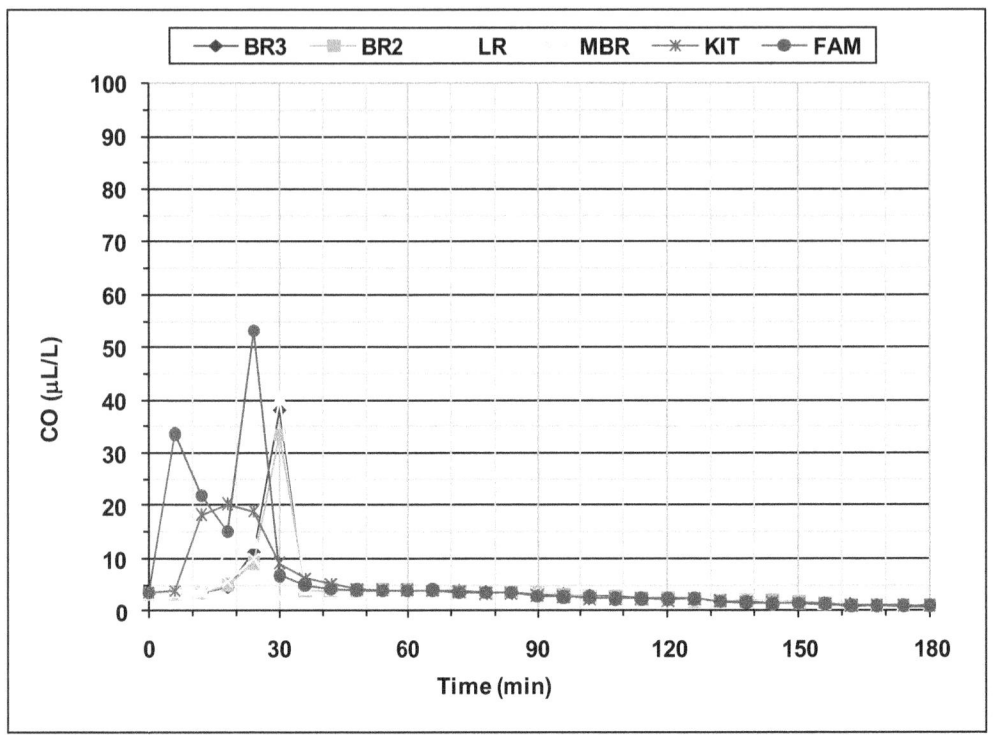

Figure 18b CO concentrations in the house for Test T (Gen SO1, Configuration 2)

As shown in Figure 18b, an initial spike of CO exceeding 50 μL/L was measured in the family room about 25 min after the generator was started, but 5 min after that it dropped below 10 μL/L and continued to drop for the remainder of the test. By comparison, unmod Gen X in Test F produced a maximum CO concentration in the family room at just over 200 μL/L.

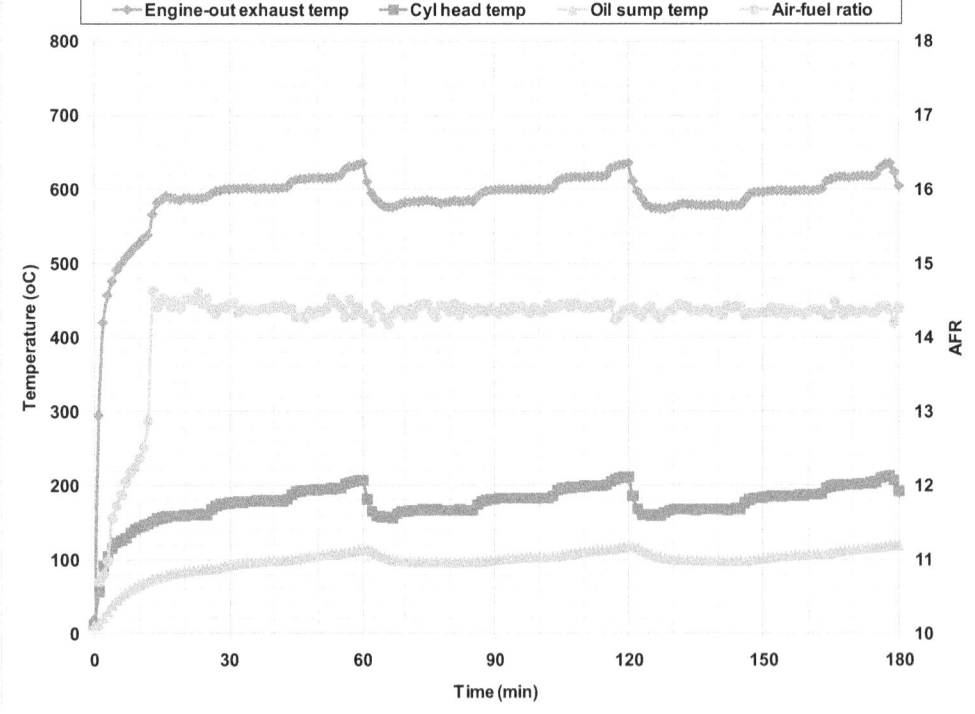

Figure 18c Temperatures and AFR measured in Test T (Gen SO1, Configuration 2)

The AFR and temperatures measured on Gen SO1 during Test T are shown in Figure 18c. The engine operated at the calibrated AFR after the engine oil temperature warmed to about 70 °C.

After this series of tests was conducted, due to limitations in the test program resources that would not support continued testing of both prototypes, a decision was made to continue the testing only the newer prototype Gen SO1 for drawing comparisons between performance of the prototype and stock generator.

Figures 19a, 19b, and 19c show the results for Test I, which was a four hour test of unmod Gen X in Configuration 3 (garage bay door closed, garage access door to house open 5 cm, and the house central HVAC fan on). These conditions are similar to the three hour Test B with unmod Gen X under Configuration 1 except for the HVAC fan status. Since the operation of HVAC fan primarily affects the airflow between rooms in the house and is not expected to significantly impact the airflow between the house and garage, this allows a comparison between the garage CO and oxygen levels in Tests I and B. After the generator was stopped, the building naturally decayed for one hour before the garage and house were mechanically vented.

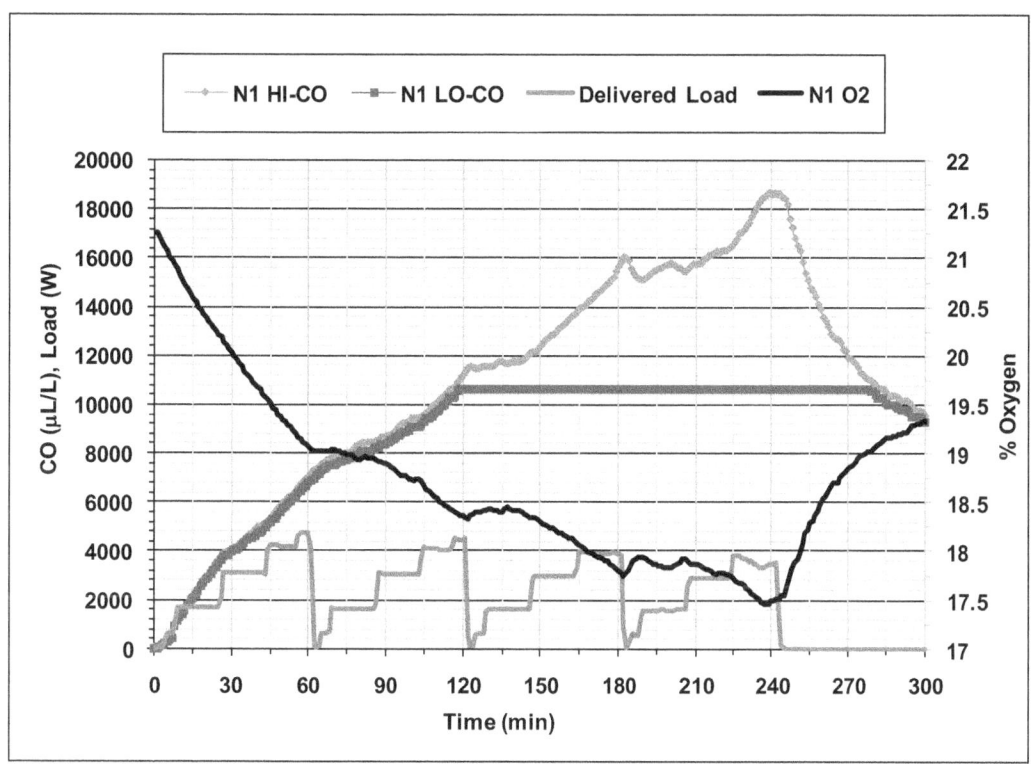

Figure 19a CO and O_2 concentrations in the garage and measured load for Test I (unmod Gen X, Configuration 3)

Figure 19a shows that the concentration of CO in the garage reached a peak of about 18,600 μL/L and the concentration of O_2 in the garage had dropped by 3.7 % to 17.5 % when the generator was stopped. It also shows that in the first load cycle the delivered electrical output was less than the load bank settings for the two highest loads in the load cycle, 4500 W and 5500 W, which were applied as the oxygen was approaching 19 %. As the oxygen continued to drop in the subsequent load cycles, the delivered power for these load points decreased further. These results are fairly similar to those in Test B.

35

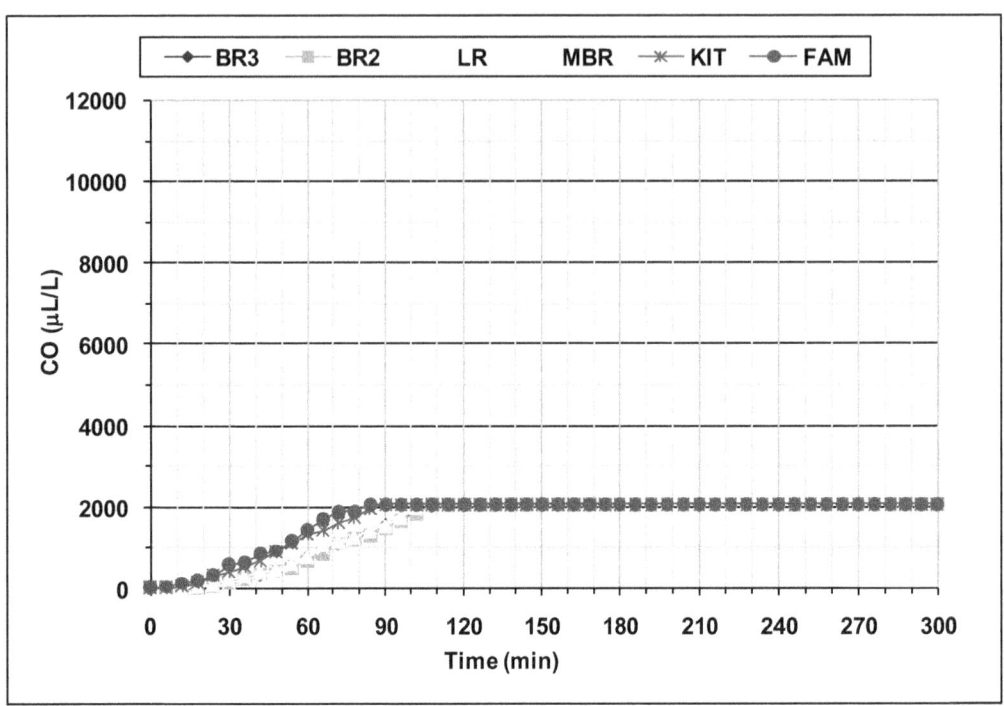

Figure 19b CO (ppm range) concentrations in the house for Test I
(unmod Gen X, Configuration 3)

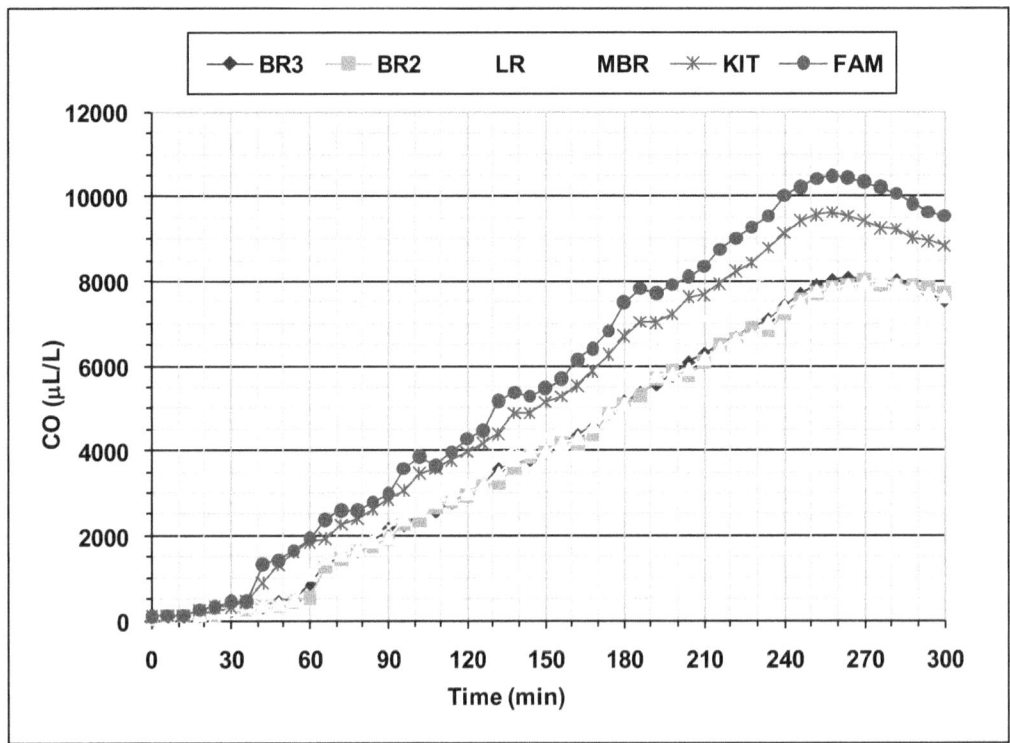

Figure 19c CO (high range) concentrations in the house for Test I
(unmod Gen X, Configuration 3)

Figures 19b and 19c show the CO concentration in the rooms of the test house, as measured in the 'ppm range' (where the CO concentration plot plateaus at the instrument's 2000 µL/L limit) and with the 'high range' CO instruments, respectively. The CO reached a peak concentration of around 10,500 µL/L in the family room, with peak concentrations in the other rooms ranging from about 8,000 µL/L to 10,000 µL/L. With the HVAC fan on in this test, there is a relatively more uniform distribution of CO compared to Test B in which the HVAC fan was off.

Figures 20a, 20b, and 20c show the results for Test Z, which was a 4.75 h test of Gen SO1 with the noncat muffler (Configuration 3). The test ended when the generator ran out of fuel. (Note: this run time does not indicate a limit on potential run-time as the tank was not full at the beginning of the test.) The test house configuration conditions are the same as that in the 4 h Test I with unmod Gen X. They are also the same as that used in the 2 h Test N with Gen SO1 except that the HVAC fan was off in Test N and Gen SO1 had the catalyst-installed muffler (referred to as catmuffler) in Test N. Since the operation of the HVAC fan is not expected to significantly impact the airflow between the house and garage, the effect of the catalytic and non-catalytic muffler on the resulting garage CO and oxygen levels between Tests Z and N (Configuration 3 and 1, respectively) up to the 2 h point can be seen. After the generator was stopped, the garage and house were mechanically vented.

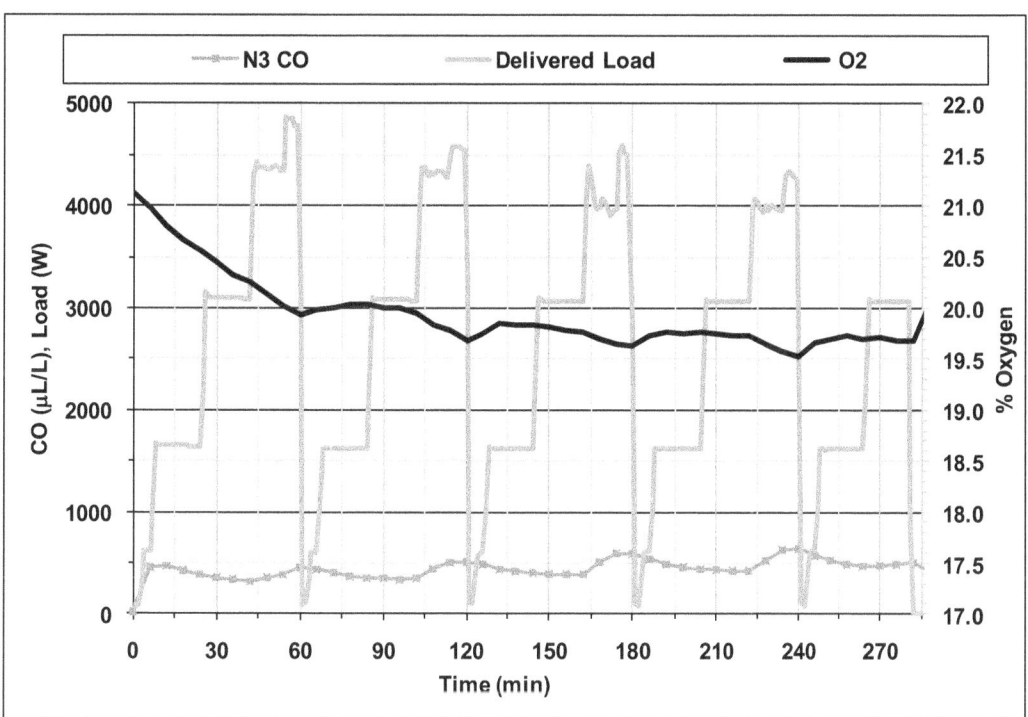

Figure 20a CO and O_2 concentrations in the garage and measured load for Test Z (Gen SO1 noncat, Configuration 3)

As shown in Figure 20a, the CO concentration in the garage initially rose to about 470 µL/L upon start, then lowered after the engine warmed up. It further increased and decreased cyclically with each successive load cycle. By the end of the fourth load cycle, it had reached a nominal peak of 630 µL/L and the oxygen dropped 1.6 % to 19.5%. This peak CO concentration is a 97 % reduction compared to that measured with unmod Gen X in Test I in which the garage CO reached about 18,600 µL/L at the end of the fourth load cycle.

Figure 20a also shows that the delivered electrical output was progressively less than the load bank settings for the two highest loads in the load cycle as the oxygen dropped throughout the test.

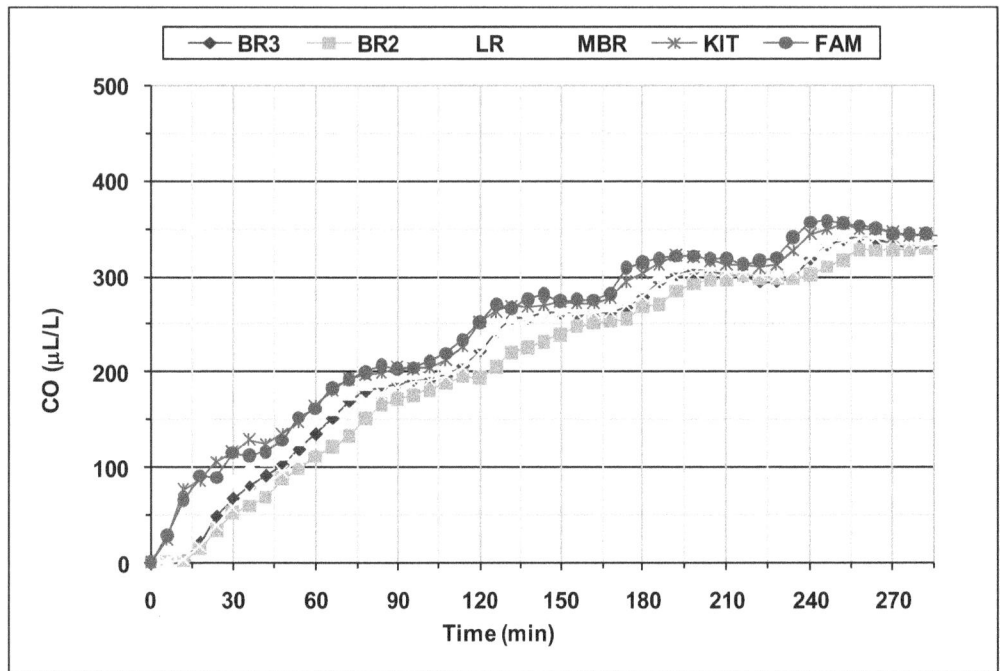

Figure 20b CO concentrations in the house for Test Z (Gen SO1 noncat, Configuration 3)

As shown in Figure 20b, the CO concentration reached a peak of nominally 360 μL/L at 4 h in the family room. There is a relatively even distribution, with all the rooms reaching at least 300 μL/L, as would be expected with the HVAC fan on. By comparison, unmod Gen X in Test I produced a peak CO concentration of around 10,600 μL/L in the family room, with peak concentrations in the other rooms ranging from about 8,000 μL/L to 10,000 μL/L.

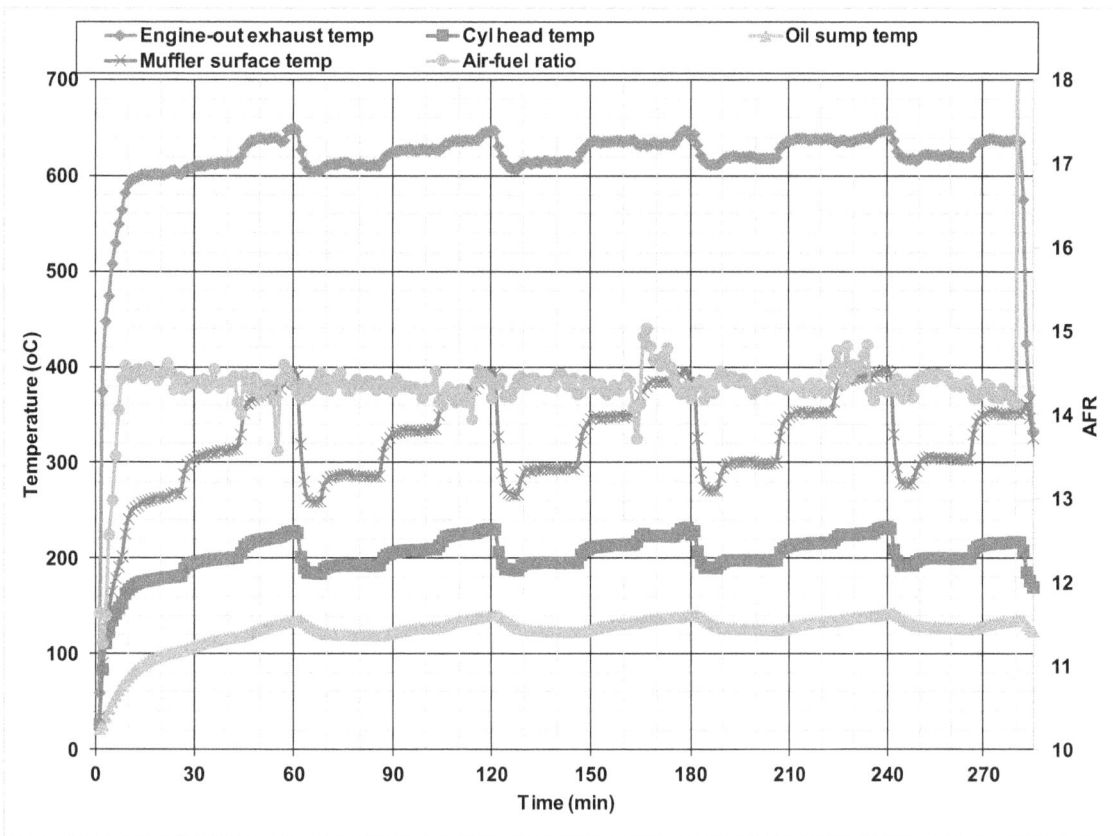

Figure 20c Temperatures and AFR measured on Gen SO1 in Test Z (Gen SO1 noncat, Configuration 3)

The AFR and temperatures measured on Gen SO1 during Test Z are shown in Figure 20c. With the exception of a few short periods of AFR excursion after the engine warmed up, the engine operated at the calibrated AFR of about 14.5. The spike in AFR at the end of the test corresponds to when the engine was turned off.

Engine operation was, by and large, comparable between Test Z and Test N, which were with Gen SO1 and the catmuffler respectively. Therefore, the garage CO concentrations at the same time in each test can be compared to get an indication of the catalyst's performance in further lowering the CO emissions. At 2 h into Test Z, the garage CO concentration was 500 μL/L and the oxygen was 19.7 %. By comparison, at the end of the 2 h Test N, the garage CO concentration reached a peak of around 300 μL/L and the garage O_2 concentration dropped to 19.4 %. Therefore, the CO concentrations were approximately 40 % lower for Test Z with the catalyst than for Test N with the EMS alone. An unknown portion of the difference may be due to differences in ambient or other test conditions.

Figures 21a and 21b show the results for the two and a quarter hour test of unmod Gen X, Test J, in Configuration 4 (garage bay door closed, garage access door to house closed, and the house central HVAC fan on). After the generator was stopped, the garage was mechanically vented. For this test, the load cycle was applied in reverse order to that shown in Table 1.

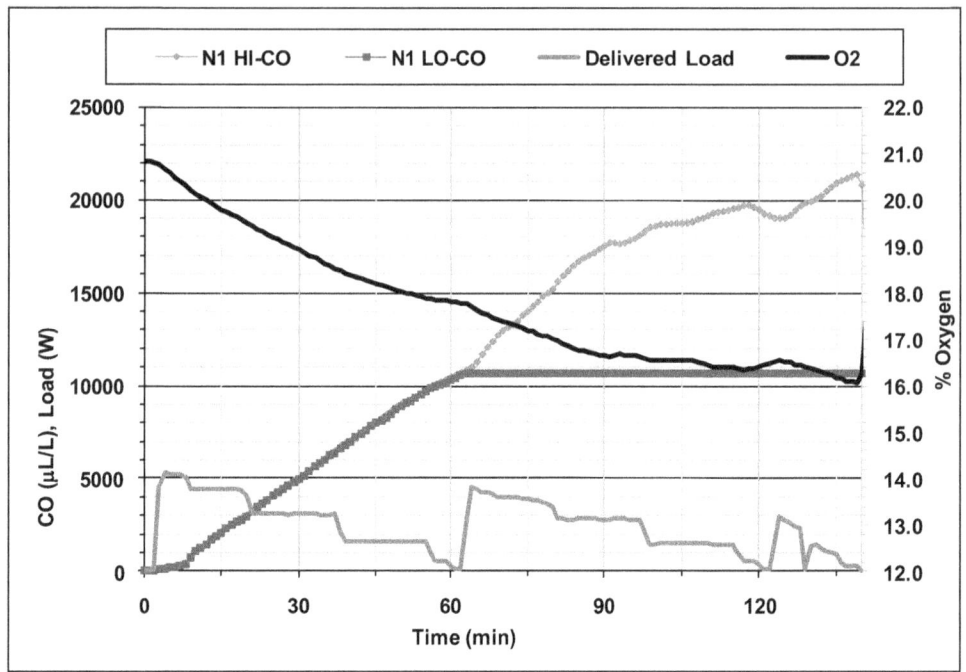

Figure 21a CO (high range) and O₂ concentrations in the garage and measured load for Test J (unmod Gen X, Configuration 4)

As shown in Figure 21a, at the time the generator was stopped, the garage CO concentration reached a peak of over 21,300 µL/L and the oxygen dropped by 4.7 % to about 16 %. It also shows that in the first load cycle, the delivered electrical output matched the load bank settings with the exception of the 5500 W setting. However, during the third load cycle, as the oxygen level dropped significantly, the generator's ability to meet the load was severely compromised and the test was ended due to poor generator operation.

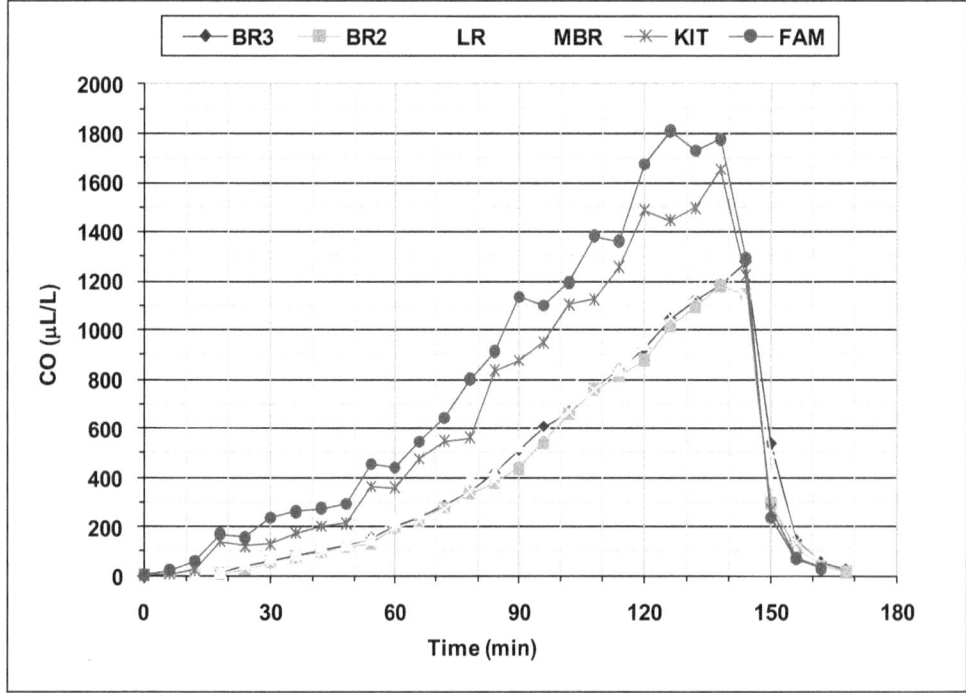

Figure 21b CO concentrations in the house for Test J (unmod Gen X, Configuration 4)

40

As shown in Figure 21b, the CO reached a peak concentration of about 1,800 μL/L in the family room, with peak concentrations in the other rooms ranging from about 1,200 μL/L to 1,650 μL/L.

Figures 22a, 22b, and 22c show the results for a six hour test of Gen SO1, Test W, with the same test house configuration as used in Test J of unmod Gen X (Configuration 4). The load cycle was applied with the same profile as that in Table 1, with the load going from low to high. After the generator was stopped, the garage was mechanically vented.

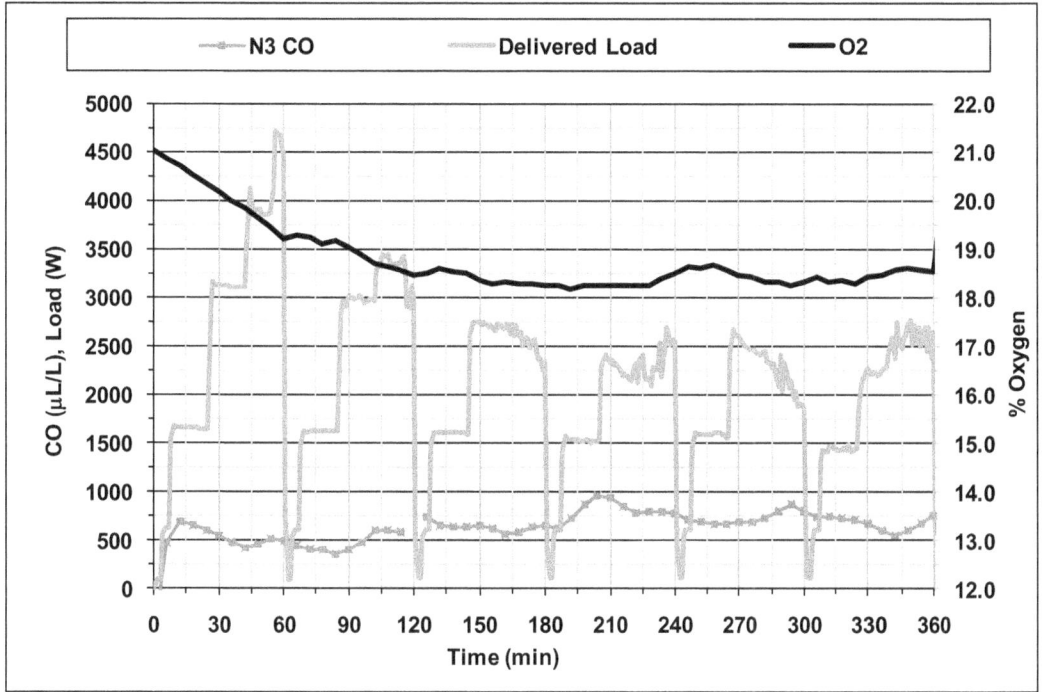

Figure 22a CO and O_2 concentrations in the garage and measured load for Test W (Gen SO1, Configuration 4)

As shown in Figure 22a, the CO concentration in the garage initially rose to 680 μL/L, and then decreased after the engine warmed up. In the fourth load cycle, it reached a peak of about 960 μL/L and the oxygen lowered by 2.8 % to 18.2 %.

At two and one quarter hours into this test, the garage CO concentration was nominally 640 μL/L. Although the tests are not entirely comparable due to the opposite loading pattern, this CO concentration is a 97 % reduction compared to that measured with unmod Gen X in Test J in which the garage CO was over 21,300 μL/L at the same time during the test.

In the first load cycle, the delivered electrical output exceeded the load bank settings except for the two highest loads. In the subsequent load cycles, as the oxygen level dropped, the delivered load was less than the load bank settings for the three highest loads in the cycle.

41

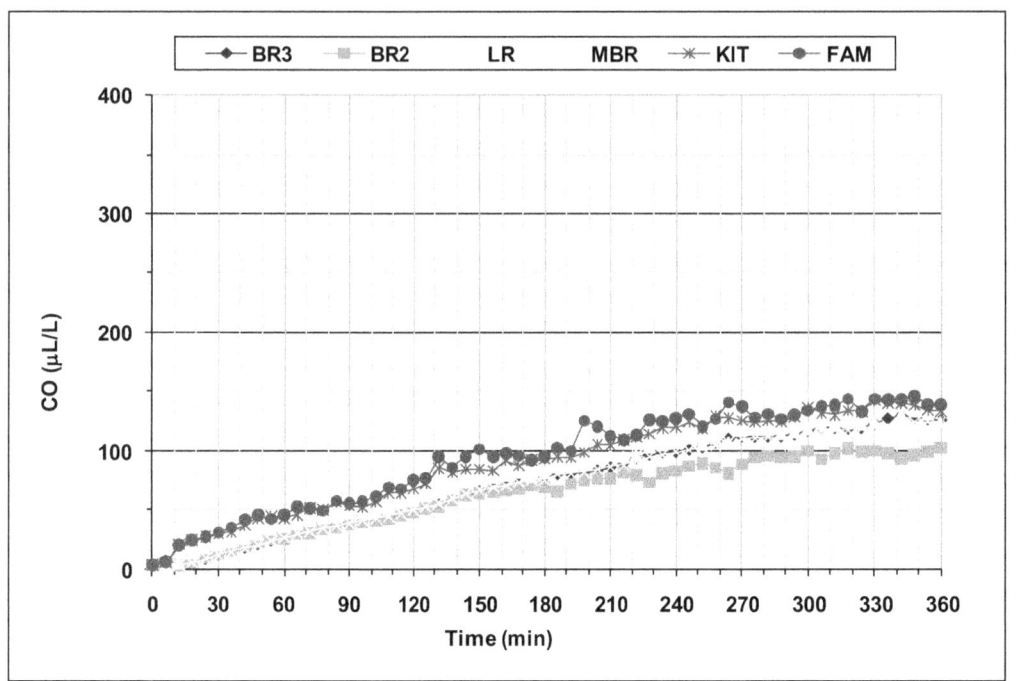

Figure 22b CO concentrations in the house for Test W (Gen SO1, Configuration 4)

As shown in Figure 22b, the CO reached a peak concentration of about 145 µL/L in the family room, with peak concentrations in the other rooms relatively evenly distributed just below that, down to 100 µL/L. By comparison, unmod Gen X in Test J produced a peak CO concentration of over 1,800 µL/L in the family room after 2 h of operation.

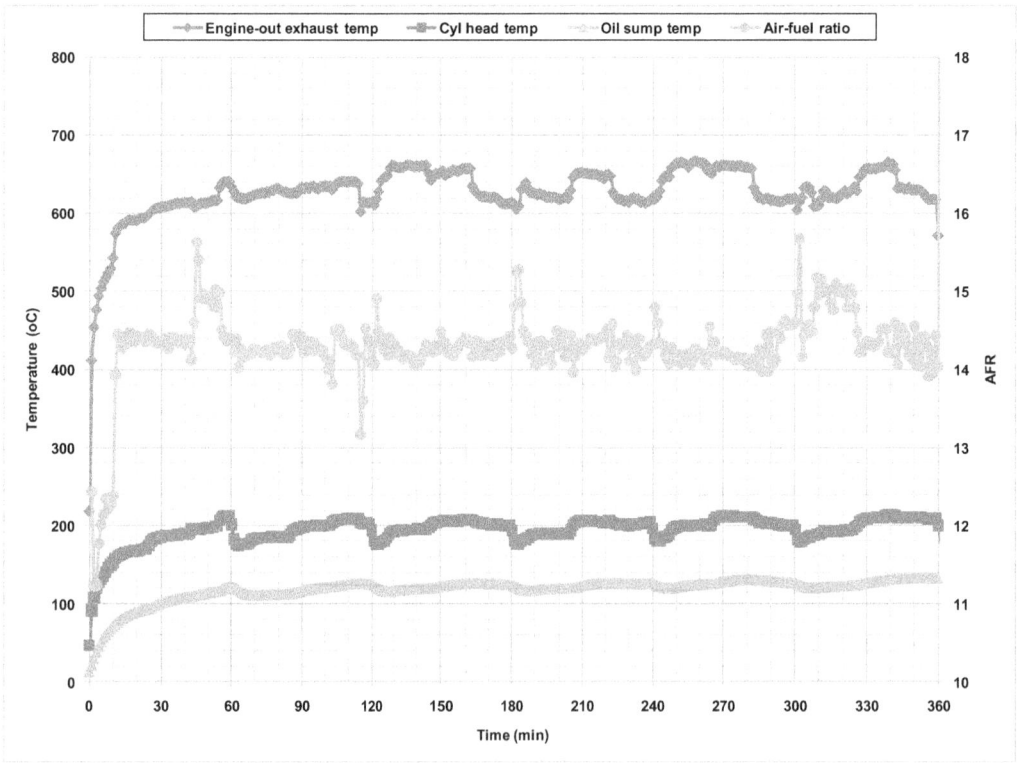

Figure 22c Temperatures and AFR measured in Test W (Gen SO1, Configuration 4)

42

The AFR and temperatures measured on Gen SO1 during Test W are shown in Figure 22c. After the engine warmed up, the engine operated at the calibrated AFR for the next 30 min. There were then occasional periods of lean as well as rich operation, with most of them occurring during the transition between the load cycles when the load bank was switched from 5500 W to no load. The spike in AFR at the end of the test corresponds to when the engine was turned off.

Figures 23a and 23b show the results for the two hour test of unmod Gen X, Test D, in Configuration 5 (garage bay door closed, garage access door to house closed, and the house central HVAC fan off). These conditions are the same as the two and a quarter hour Test J with unmod Gen X except in that test the HVAC fan was on. Since the operation of the HVAC fan is not expected to significantly impact the airflow between the house and garage, especially with the house door closed, this allows some degree of comparison to be made for the resulting garage CO and oxygen levels between Tests D and J. After the generator was stopped, the garage was mechanically vented.

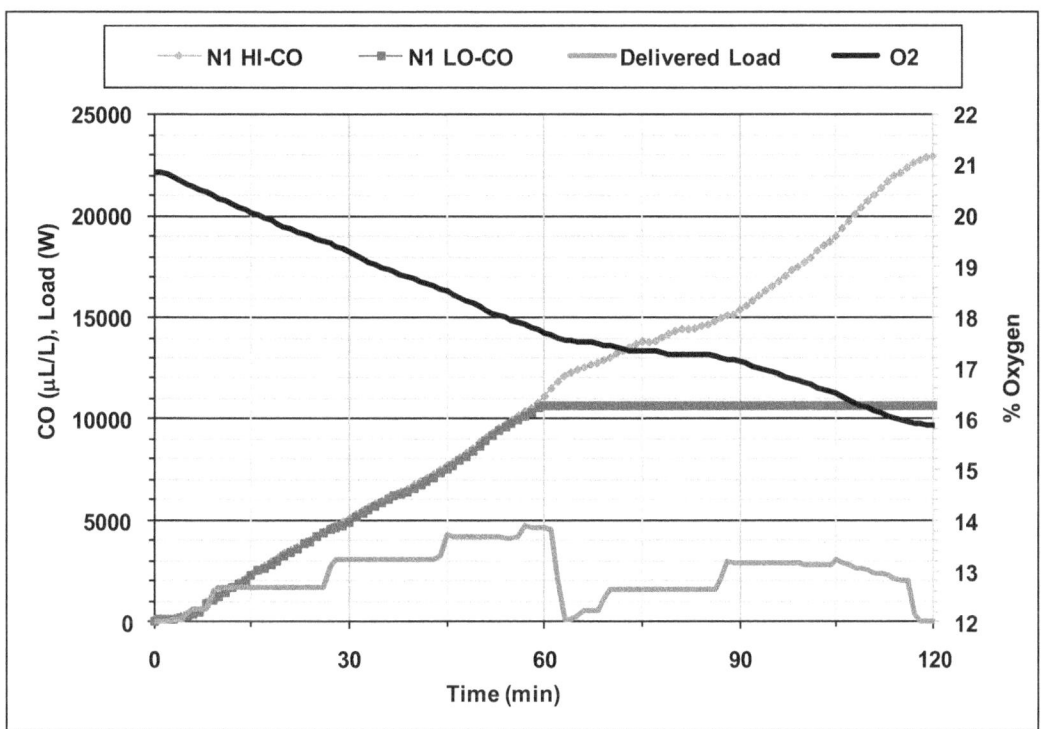

Figure 23a CO (high range) and O_2 concentrations in the garage and measured load for Test D (unmod Gen X, Configuration 5)

Figure 23a shows the concentration of CO in the garage reached a peak of almost 23,000 µL/L, and the concentration of O_2 in the garage dropped by 5.0 % to below 16 % when the generator was stopped. It also shows that in the first load cycle the delivered electrical output was less than the load bank settings for the two highest loads in the load cycle, 4500 W and 5500 W, which were applied as the oxygen was approaching 18 %. As the oxygen continued to drop in the subsequent load cycle, the delivered power for these load points decreased further. The results are similar to those in Test J despite the reversal of the load cycled pattern in Test J.

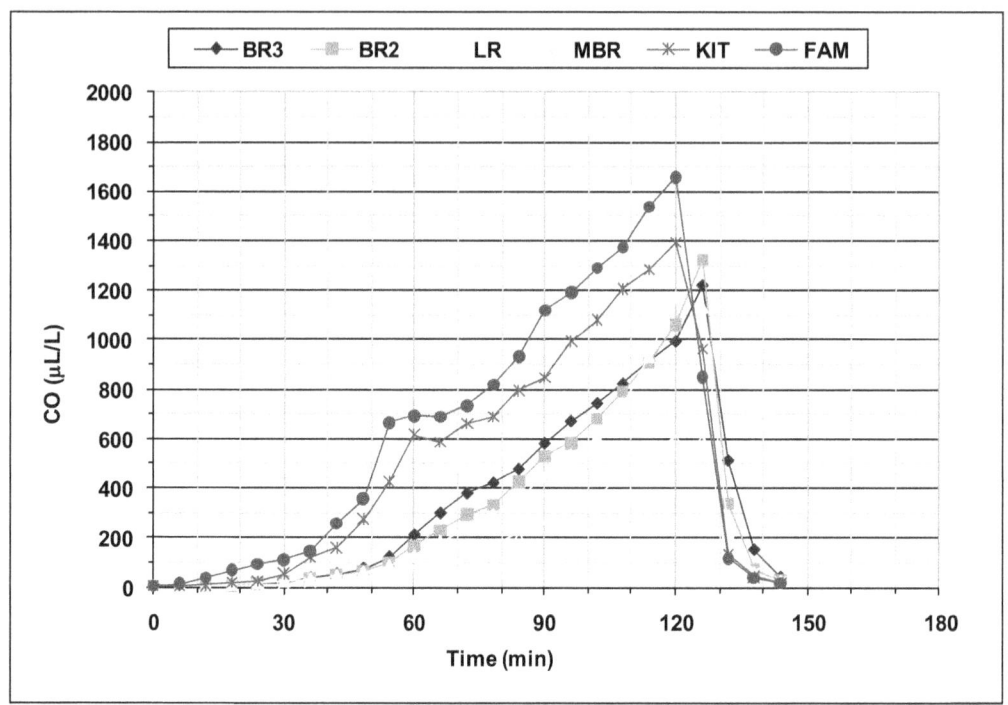

Figure 23b CO concentrations in the house for Test D (unmod Gen X, Configuration 5)

Figure 23b shows the CO reached a peak concentration of almost 1660 μL/L in the family room with peak concentrations in the other rooms ranging from about 600 μL/L to 1400 μL/L. This is a comparable peak CO concentration to the 1670 μL/L measured in the family room at the 2 h point in Test J. When comparing the other room time course profiles with those at the 2 h point in Test J, it can be observed that the mixing due to the operation of the HVAC fan made the most difference in the master bedroom. This effect is not consistent during all tests as other factors affecting mixing (such as temperature) differ from test to test.

Figures 24a, 24b, and 24c show the results for Test AH, which was a five hour test of Gen SO1 with the noncat muffler and the same conditions of the test house as used in the 2 h Test D with unmod Gen X (Configuration 5). These conditions are also the same as that used in the 6 h Test W with Gen SO1 except that in Test W Gen SO1 had the catmuffler and the HVAC fan was on. Since the operation of the HVAC fan is not expected to significantly impact the airflow between the house and garage, this allows some degree of comparison to be made for the resulting garage CO and oxygen levels between Tests AH and W. After the generator was stopped, the exhaust decayed naturally for 45 min and then the garage and house were mechanically vented.

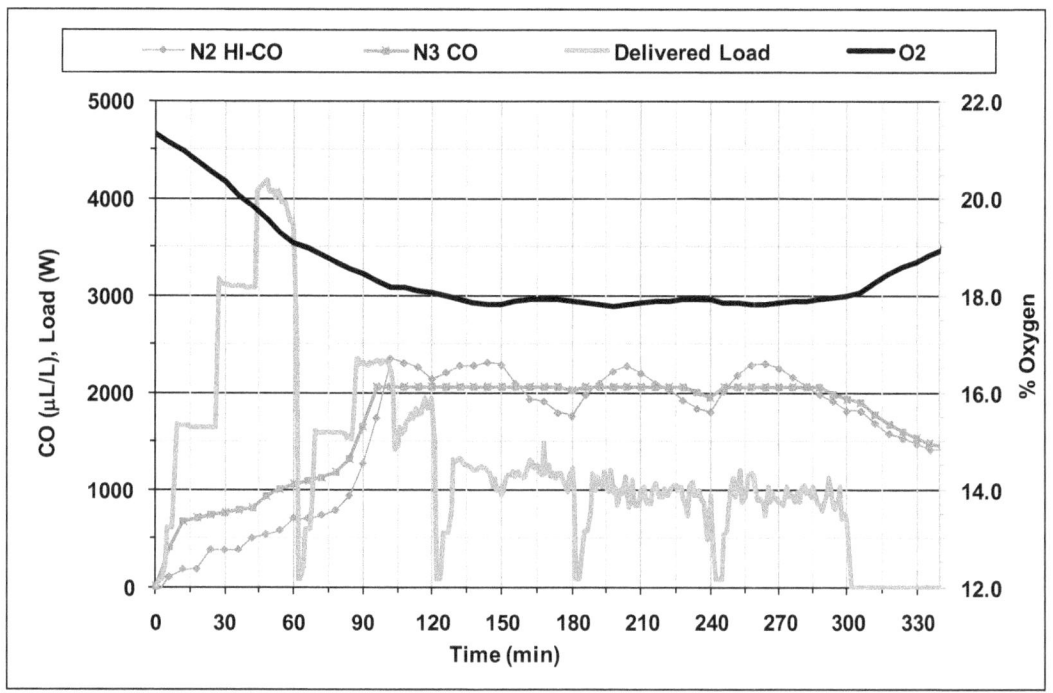

Figure 24a CO and O_2 concentrations in the garage and measured load for Test AH (Gen SO1 noncat, Configuration 5)

As shown in Figure 24a, the CO concentration in the garage initially rose to nominally 670 µL/L upon start, then continued to climb until it reached a nominal peak of 2300 µL/L and oxygen lowered 3.5 % to 17.8 % in the garage during the second load cycle. This CO concentration is a 90 % reduction compared to that measured with unmod Gen X in Test D in which the CO in the garage at the end of the second load cycle was almost 23,000 µL/L.

Figure 24a also shows that in the first load cycle the delivered electrical output was less than the load bank settings for the two highest loads in the load cycle. During the subsequent load cycles the delivered power degraded even further as the garage oxygen approached and then dropped below 18 %.

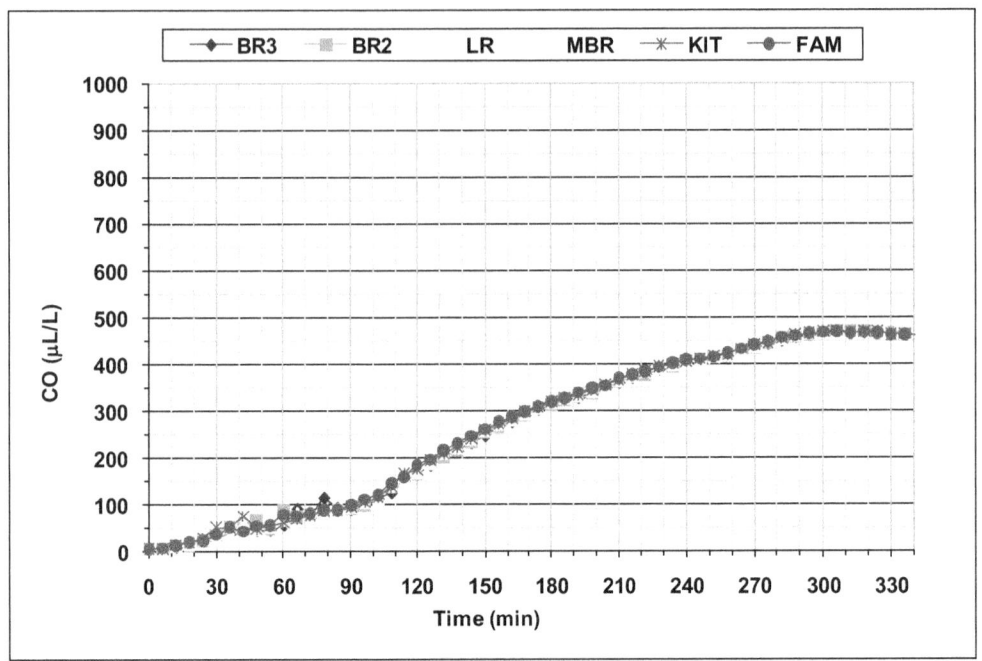

Figure 24b CO concentrations in the house for Test AH (Gen SO1 noncat, Configuration 5)

As shown in Figure 24b, the CO reached a peak concentration of about 470 μL/L throughout the house, with even distribution among the rooms even though the HVAC fan was off. At 2 h into this test, the CO in the house was about 180 μL/L. By comparison, the 2 h operation of unmod Gen X in Test D produced a peak CO concentration of almost 1660 μL/L in the family room, with peak concentrations in the other rooms ranging from about 600 μL/L to 1400 μL/L.

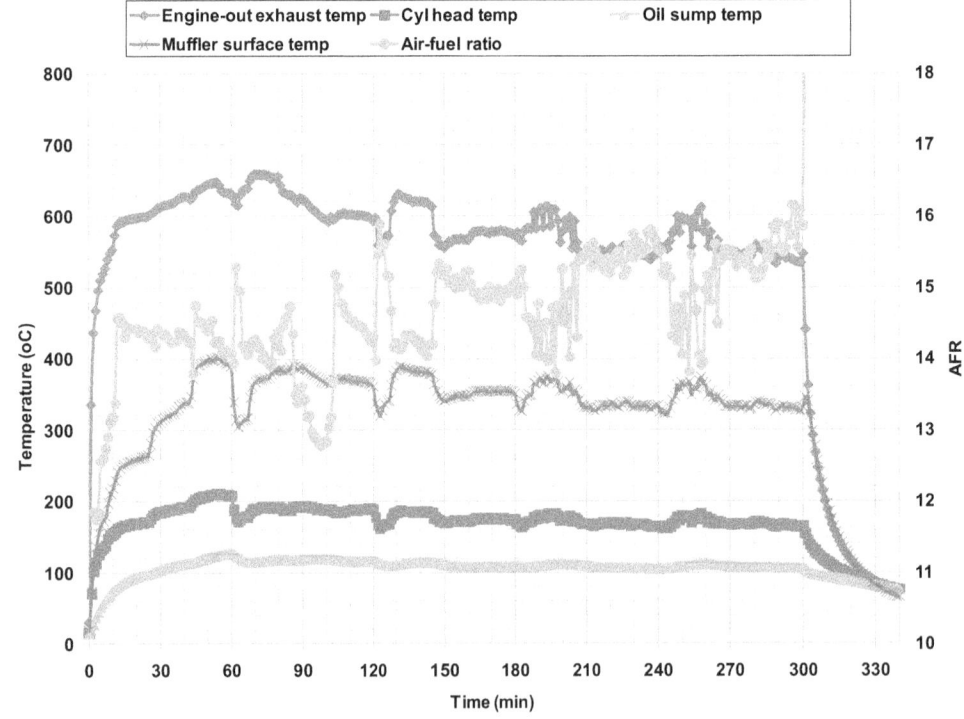

Figure 24c Temperatures and AFR measured in Test AH (Gen SO1 noncat, Configuration 5)

The AFR and temperatures measured on noncat Gen SO1 during Test AH are shown in Figure 24c. After the engine warmed up, it operated at the calibrated AFR for the next 30 min, but then had periods of off-design operation throughout the remainder of the test. The spike in AFR at the end of the test corresponds to when the engine was turned off.

Since engine performance during the first 40 min in Test AH was similar to that in Test W with Gen SO1 and the catmuffler, a comparison of each test's garage CO concentration at that point in time suggests the prototype's catalyst is providing about a 50 % reduction in the CO emissions compared with that provided by the EMS alone. At 40 min, the garage CO concentrations were about 410 μL/L and 820 μL/L in Tests W and AH, respectively. This reduction is somewhat larger than the 40 % reduction observed when comparing the garage CO concentrations in Tests Z and N.

Figures 25a and 25b show the results for Test G, a 2 h test of unmod Gen X in Configuration 6 (garage bay door open, garage access door to house open 5 cm, and the house central HVAC fan on). After the generator was stopped, the garage was mechanically vented.

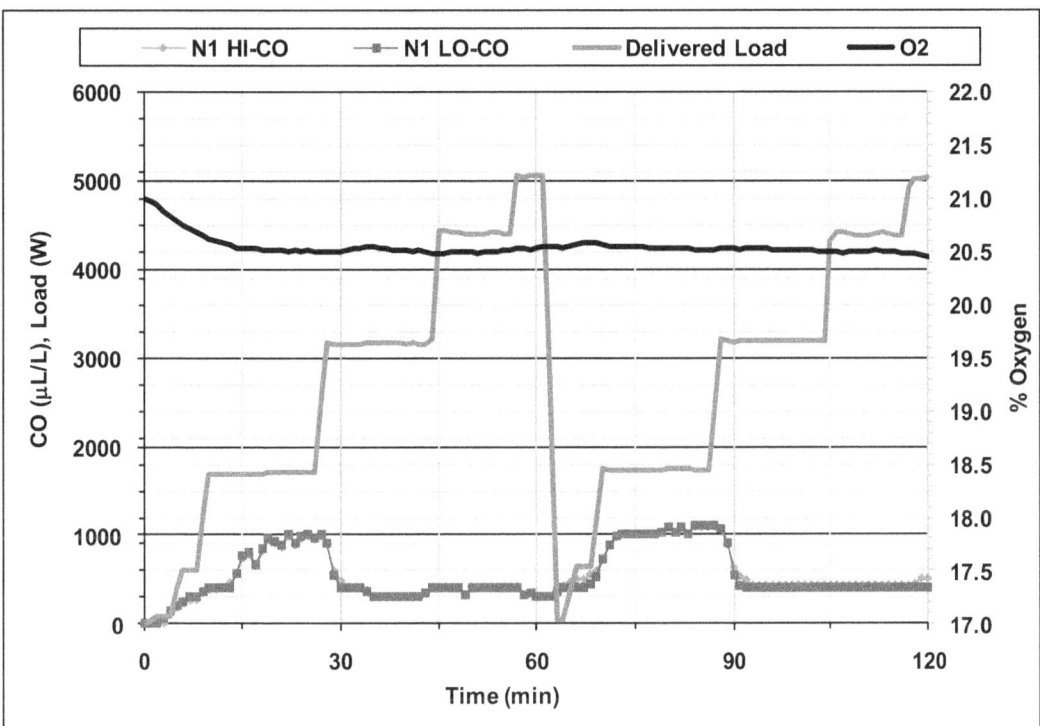

Figure 25a CO and O_2 concentrations in the garage and measured load for Test G (unmod Gen X, Configuration 6)

As shown in Figure 25a, the CO in the garage peaked at around 1100 μL/L during the second load cycle (though the instrument uncertainty is large relative to the concentrations). With the garage bay door open, the oxygen level decreased 0.5 % to about 20.5 %. Throughout the test, the delivered electrical output met or exceeded the load bank settings.

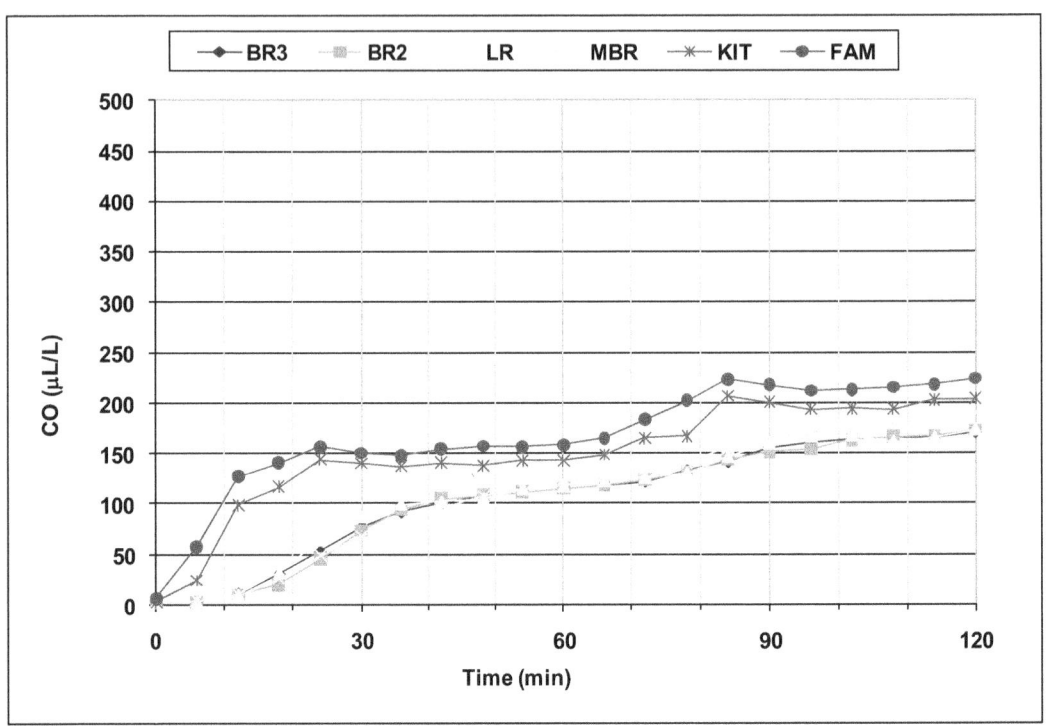

Figure 25b CO (ppm range) concentrations in the house for Test G (unmod Gen X, Configuration 6)

Figure 25b shows the CO reached a peak concentration of about 220 μL/L in the family room, with slightly lower peak concentrations in the other rooms of around 190 μL/L to 200 μL/L.

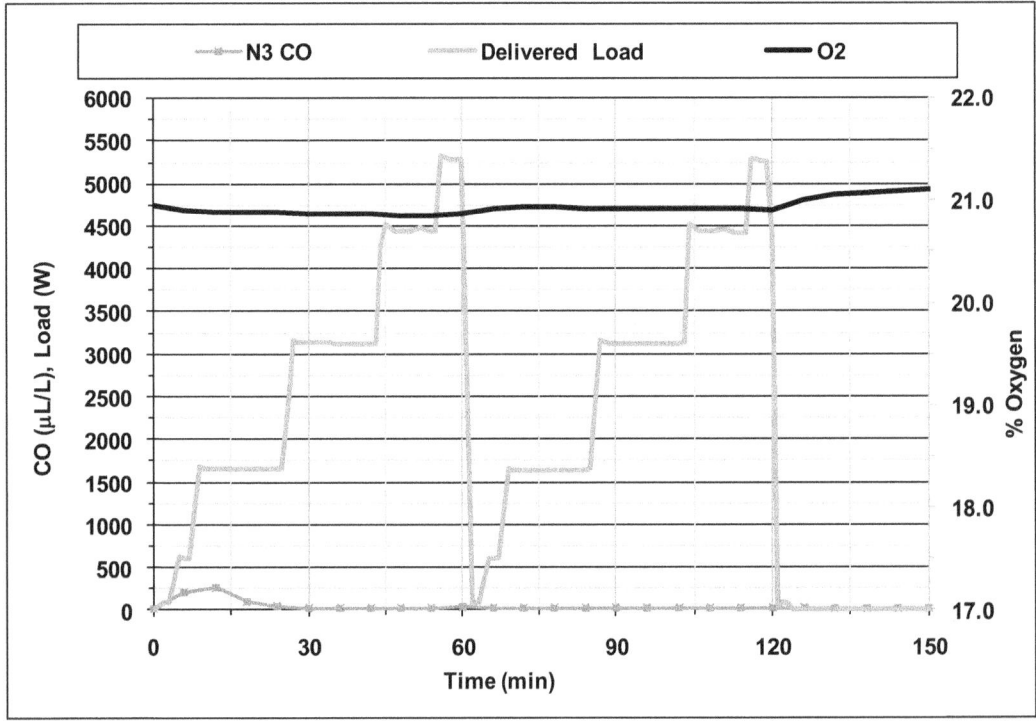

Figure 26a CO and O_2 concentrations in the garage and measured load for Test U (Gen SO1, Configuration 6)

Figures 26a, 26b, and 26c show the results for Test U, which was a 2 h test of Gen SO1 with the same conditions of the test house as used in the 2 h Test G with unmod Gen X (Configuration 6). After the generator was stopped, the exhaust decayed naturally for 30 min and then the garage and house were mechanically vented.

As shown in Figure 26a, after an initial spike to nominally 260 µL/L of CO in the garage shortly after the generator was started, it dropped and maintained a level below 30 µL/L throughout the test. After the initial spike, this CO concentration reflects about a 97 % reduction compared to that measured with unmod Gen X in Test G in which the CO in the garage ranged from around 300 µL/L to 1100 µL/L for portions of the second load cycle. With the garage bay door open, the oxygen level stayed nominally at ambient.

Throughout the test, the delivered electrical output met or exceeded the load bank settings with the exception of the highest load setting.

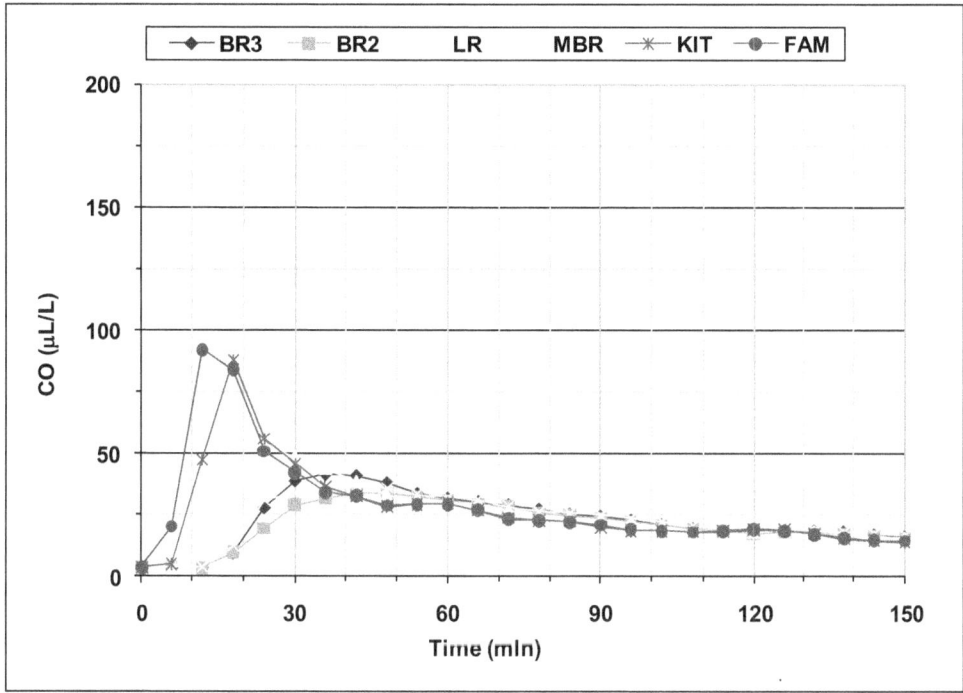

Figure 26b CO concentrations in the house for Test U (Gen SO1, Configuration 6)

As shown in Figure 26b, the CO concentration in the family room initially spiked to about 90 µL/L and then dropped to an even distribution in all rooms of the house around 30 µL/L with a continual decline to below 20 µL/L before mechanical venting was initiated. By comparison, unmod Gen X in Test G produced a nominal peak CO concentration of 220 µL/L in the family room with a minimum peak concentration in the other rooms just below 190 µL/L.

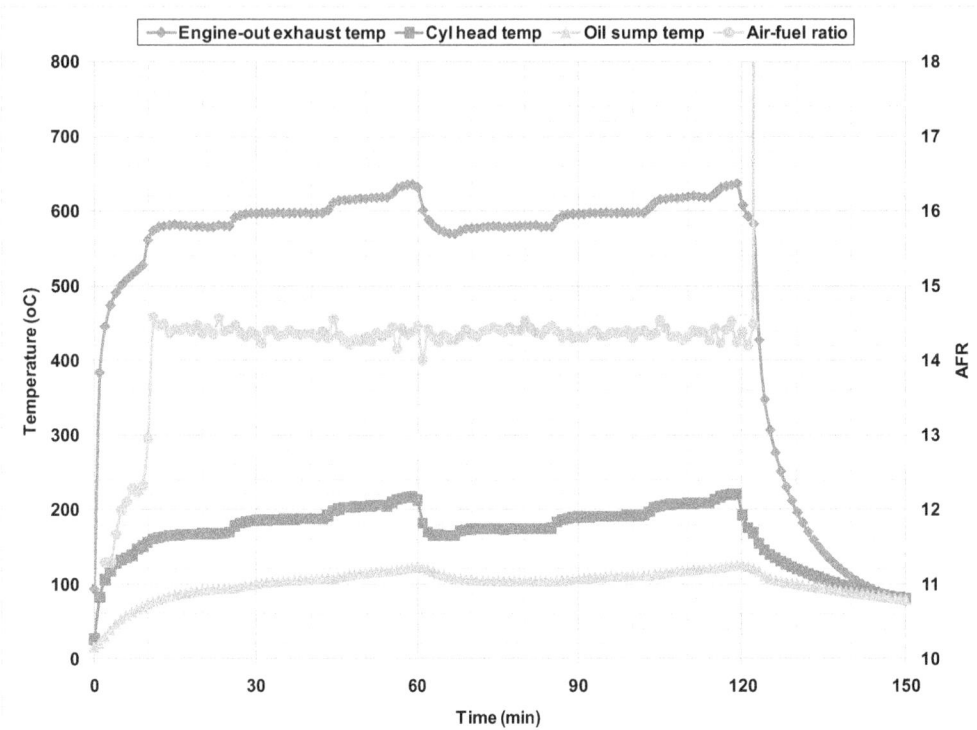

Figure 26c Temperatures and AFR measured in Test U (Gen SO1, Configuration 6)

The AFR and temperatures measured on Gen SO1 during Test U are shown in Figure 26c. The engine operated at the calibrated AFR after the engine oil temperature warmed to nominally 70 $^\circ$C. The spike in AFR at the end of the test corresponds to when the engine was turned off.

Figures 27a and 27b show the results for Test K, which was a 2 h 10 min test of unmod Gen X in Configuration 7 (garage bay door and garage access door to house open, and house central HVAC fan off). For this test, the load cycle was applied in reverse order to that shown in Table 1. The test house conditions for this test are similar to the 2 h Test G with unmod Gen X except in that test the HVAC fan was on. Since the operation of the HVAC fan is not expected to have a significant impact on the airflow between the house and garage, this allows some degree of comparison to be made for the resulting garage CO and oxygen levels between Tests K and G. After the generator was manually stopped, the garage and house were mechanically vented.

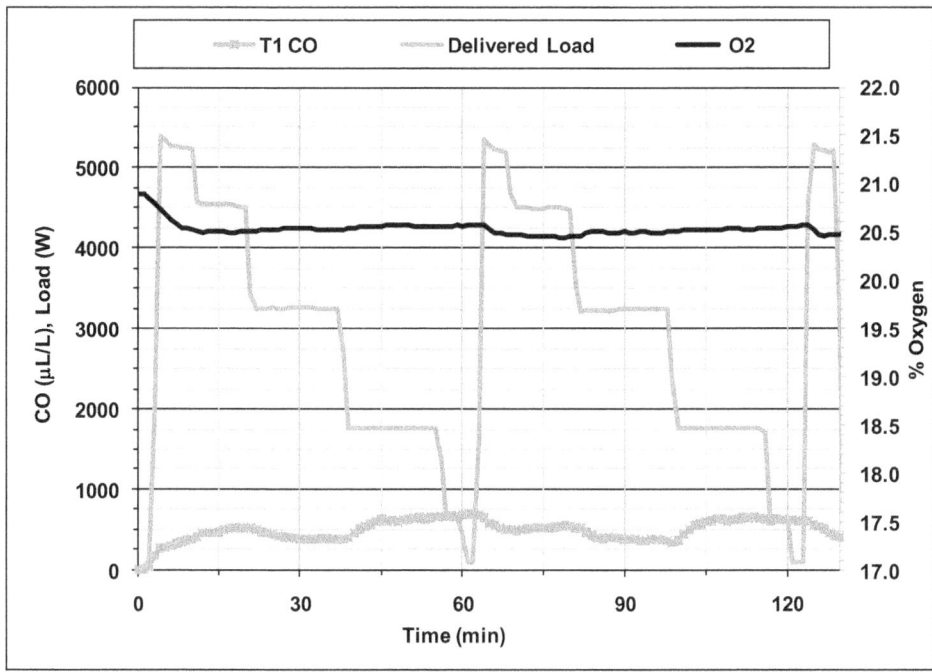

Figure 27a CO and O_2 concentrations in the garage and measured load for Test K (unmod Gen X Configuration 7)

As shown in Figure 27a, the CO in the garage peaked at about 680 μL/L. This compares to the 1100 μL/L reported in Test G with unmod Gen X that was measured with a high range CO analyzer. With the garage bay door open, the garage oxygen level decreased to about 20.4 %. Throughout the test, the delivered electrical output exceeded the load bank settings with the exception of the highest load setting.

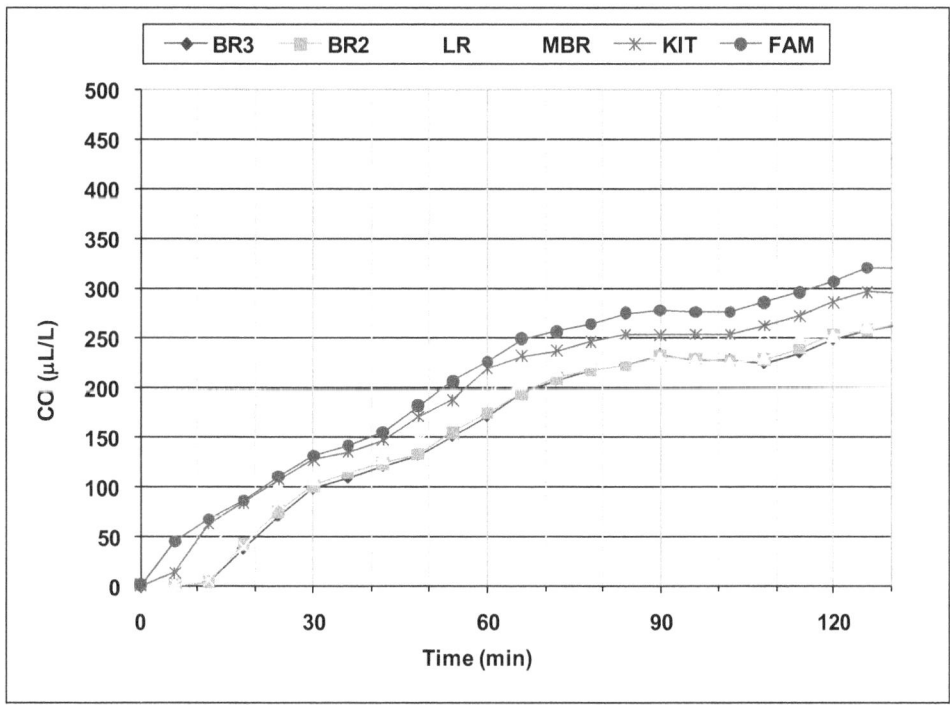

Figure 27b CO concentrations in the house for Test K (unmod Gen X Configuration 7)

Figure 27b shows the CO reached a peak concentration of 320 μL/L in the family room with peak concentrations in the other rooms just below that value, nominally 260 μL/L when mechanical venting was initiated.

Figures 28a, 28b, and 28c show the results for Test V, which was a 2 h 15 min test of Gen SO1 with the noncat muffler and the same test house configuration as used in the 2 h Test K with unmod Gen X (Configuration 7). To match the reverse order load profile used Test K, the load cycle for this test was also applied in reverse order to that shown in Table 1. The test house conditions for this test are also the same as that used in the 2 h Test U with Gen SO1 except that in Test U Gen SO1 had the catmuffler and the house central HVAC fan was on. After the generator was stopped, the garage and house were mechanically vented. Due to a software error, about 15 min of data were not recorded approximately 1 h into the test.

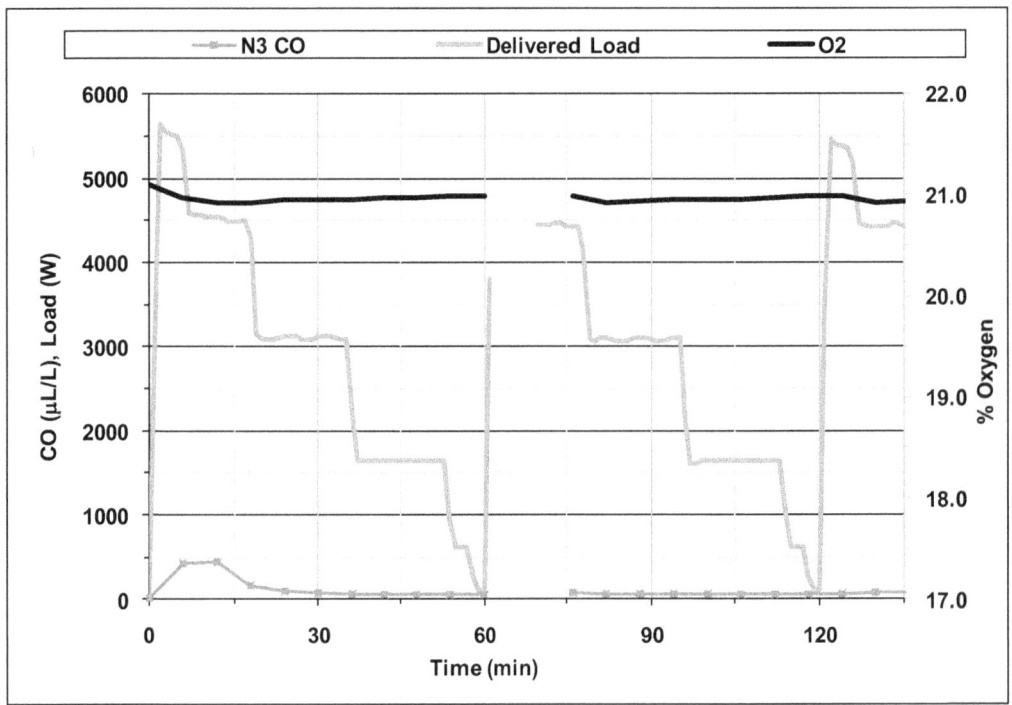

Figure 28a CO and O_2 concentrations in the garage and measured load for Test V (noncat Gen SO1, Configuration 7)

As shown in Figure 28a, after an initial spike to nominally 430 μL/L of CO in the garage shortly after the generator was started, it dropped to a level near 50 μL/L before rising to about 80 μL/L during the brief third load cycle. Note that the missing data included the high load portion of the second load cycle and a peak during this time cannot be ruled out. Excluding the initial peak of Test V, this is a reduction of 85 % to 88 % compared to that measured with unmod Gen X in Test K, in which the CO in the garage ranged from 350 μL/L to 650 μL/L.

With the garage bay door open, the garage oxygen level stayed nominally at ambient. Throughout the test, the delivered electrical output met or exceeded the load bank settings, with the exception of a slight drop at the highest setting during the third load cycle.

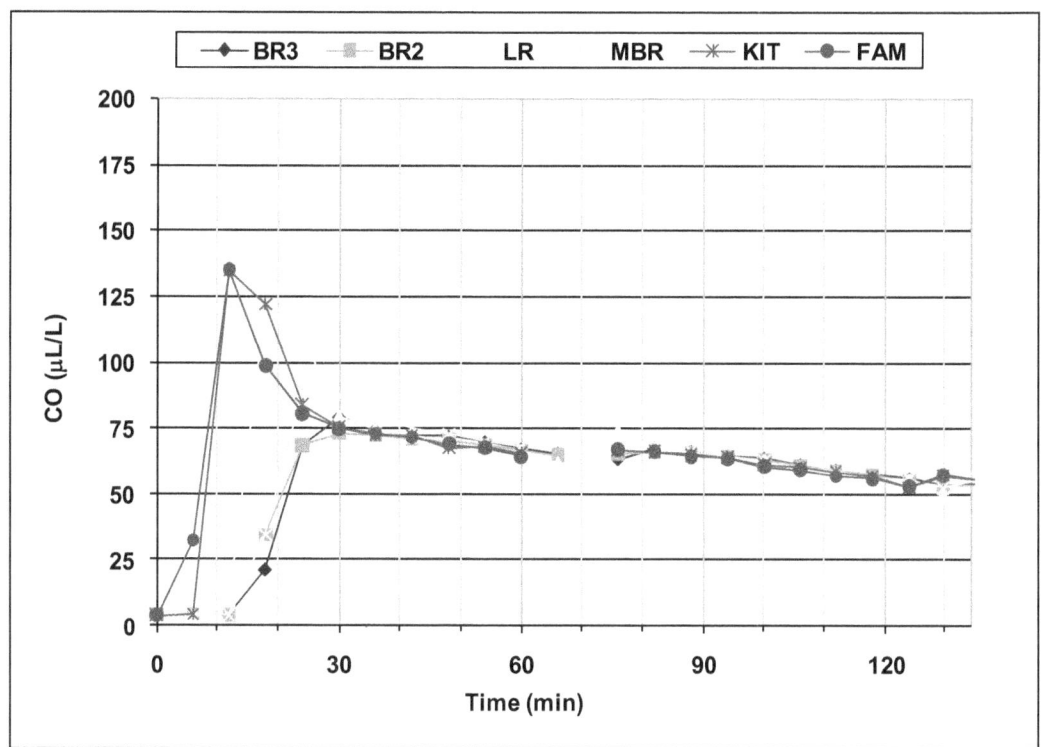

Figure 28b CO concentrations in the house for Test V (noncat Gen SO1, Configuration 7)

As shown in Figure 28b, the CO concentration in the family room initially spiked to 135 μL/L and then dropped to a uniform concentration throughout the house of around 75 μL/L, with a continual decline to 50 μL/L, at which point the mechanical venting was initiated. With the exception of the first 25 min of the test, the distribution was very uniform despite the HVAC fan being off. By comparison, unmod Gen X in Test K produced a less uniform house distribution, with a peak CO concentration of nominally 320 μL/L in the family room and concentrations in the other rooms just below that, down to 260 μL/L.

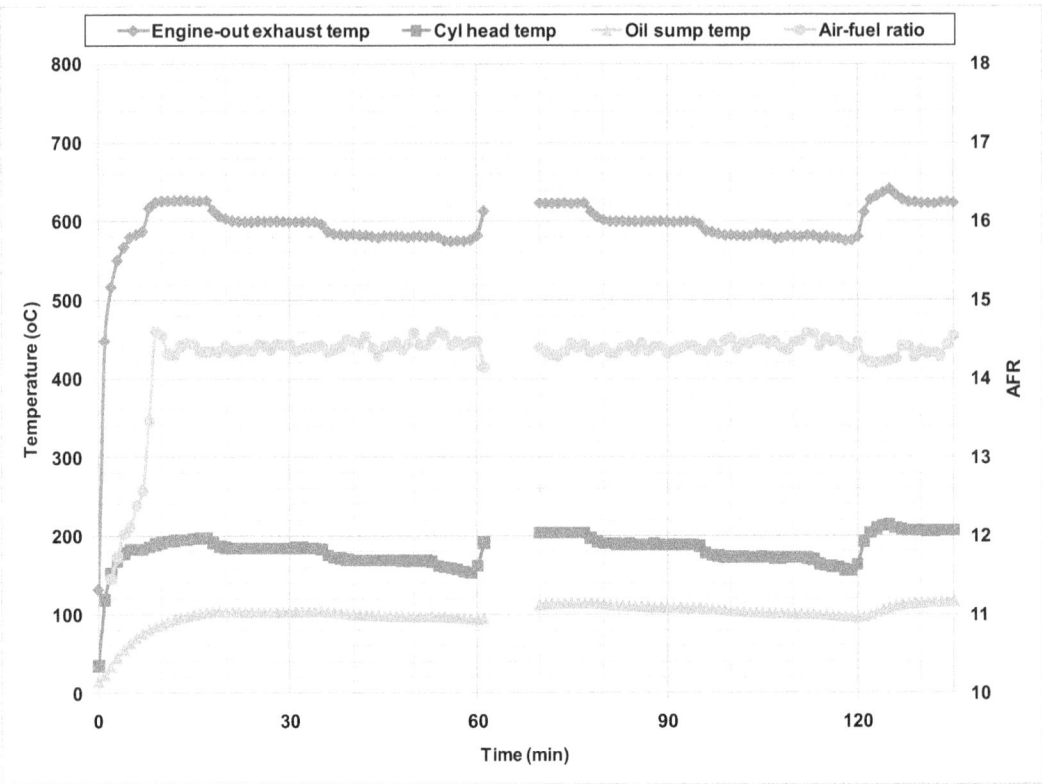

Figure 28c Temperatures and AFR measured in Test V (noncat Gen SO1, Configuration 7)

The AFR and temperatures measured on Gen SO1 during Test V are shown in Figure 28c. The engine operated at the calibrated AFR after the engine oil temperature warmed to about 70 $^\circ$C.

Engine performance in this test was similar to that in Test U with Gen SO1 and the catmuffler, with the caveats that the loads were applied in opposite order and some data were missed in Test V. A comparison of the 50 μL/L garage CO concentration in this test with the 20 μL/L in Test U indicates the prototype's catalyst is providing about up to a 60 % reduction in CO emissions from that provided by the EMS alone. This somewhat larger difference than found when comparing Tests Z to N and Tests AH to W could be due to changes in infiltration rates or other factors.

Summary of Garage Tests

A series of tests were conducted in which portable gasoline-powered electric generators were operated in the attached garage of the NIST manufactured test house. The data include CO emission and O_2 depletion in the garage, CO migration into the test house and engine operation parameters. A summary of the test results is provided in Table 4. These tests document reductions of 85 % to 98 % in CO concentrations due to emissions from two modified, prototype low CO-emission portable generators compared to an unmodified generator. The second prototype (Gen SO1) resulted in lower CO concentrations during similar tests with the garage bay door closed while both prototypes resulted in low CO concentrations during tests with the garage bay door open. Note that these results apply to the specific units tested and that other units, modifications, and test conditions may produce different results.

Table 4 Summary of Garage Test Results

Generator	Test ID	Garage bay door, house door, HVAC	Test Duration (h)	Peak Garage CO Concentration (μL/L)	% Reduction in peak garage CO relative to unmod GenX	Peak CO concentration in house (μL/L)
unmod GenX	B	Closed, open, off	3	19,500 (12,800 at 2 h)	NA	6500
modGenX	O	Closed, open, off	4.5	3000 (1,400 at 3 h)	93	800
SO1	N	Closed, open, off	2	300	98	140
unmod GenX	F	Open, closed, off	4	1,500	NA	200
modGenX	R	Open, closed, off	4	30	98	5
SO1	T	Open, closed, off	3	300 (20 after initial spike)	98	50
unmod GenX	I	Closed, open, on	4	18,600	NA	10,600
SO1 with noncat muffler	Z	Closed, open, on	4.75	630	97	360
unmod GenX	J	Closed, closed, on	2.25	21,300	NA	1,800
SO1	W	Closed, closed, on	6	960 (640 at 2.25 h)	97	145
unmod GenX	D	Closed, closed, off	2	23,000	NA	1660
SO1 with noncat muffler	AH	Closed, closed, off	5	2,300	90	470
unmod GenX	G	Open, open, on	2	1,100	NA	220
SO1	U	Open, open, on	2	260 (< 30 after initial spike)	97	90
unmod GenX	K	Open, open, off	>2	680	NA	320
SO1 with noncat muffler	V	Open, open, off	>2	430 (50 to 80 after initial spike)	85 to 88	135

Notes:

Unmod Gen X is an unmodified (stock) generator with a carbureted engine.

Mod Gen X is a modified (prototype) generator with electronic fuel injection, an engine control unit and a catalytic converter.

Gen SO1 is a modified (prototype) generator with electronic fuel injection, an engine control unit (different than mod Gen X), and a catalytic converter (not used in 'noncat' configuration).

% reduction in peak garage CO concentration excludes initial spike.

Simulation

This section of the report describes the CONTAM model of the test house used to extend the measurement results to other scenarios, e.g. weather conditions and house configuration. The primary objective of these modeling efforts was to examine the potential performance of the prototype generators under a wider range of conditions than studied during the experiments at the test house. This section starts with a detailed description of the CONTAM model of the test house, then describes a number of model validation cases, and finally presents simulation results for a broad range of cases. The purpose of the model validation effort was to establish the ability of the CONTAM model to predict CO levels in the garage and the house and to develop an estimate of the uncertainty of these predictions relative to the measured values.

Modeling Method

The simulations in this study used CONTAM (Walton and Dols 2010), a multizone IAQ and ventilation model developed in the Engineering Laboratory (EL) at NIST. The multizone approach is implemented by constructing a building model as a network of elements describing the flow paths (HVAC ducts, doors, windows, cracks, etc.) between the zones (primarily rooms) of a building. The network nodes represent the zones, which are modeled with a hydrostatically varying pressure, and uniform temperature and pollutant concentration within each zone. After calculating the airflow between zones and the outdoors, zone pollutant concentrations are calculated by applying mass balance equations to the zones. CONTAM has frequently been used to study a variety of residential IAQ issues in past simulation studies (Emmerich and Persily 1996, Emmerich et al. 2005). For more detail on the CONTAM model, see the user's manual which is available online along with the latest version of the CONTAM program (www.bfrl.nist.gov/IAQanalysis).

The graphical representation of the main floor of the test house as it appears in CONTAM is shown in Figure 29. The test house crawlspace and attic were included in the model but are not shown. The layout of the test house within CONTAM and the division of the zones were defined to represent the actual floorplan of the test house as seen in Figure 2, therefore each zone typically represents one room although the kitchen, living, dining and family room spaces are combined into a single CONTAM zone, referred to as zone LFK, due to the lack of any partitions separating them.

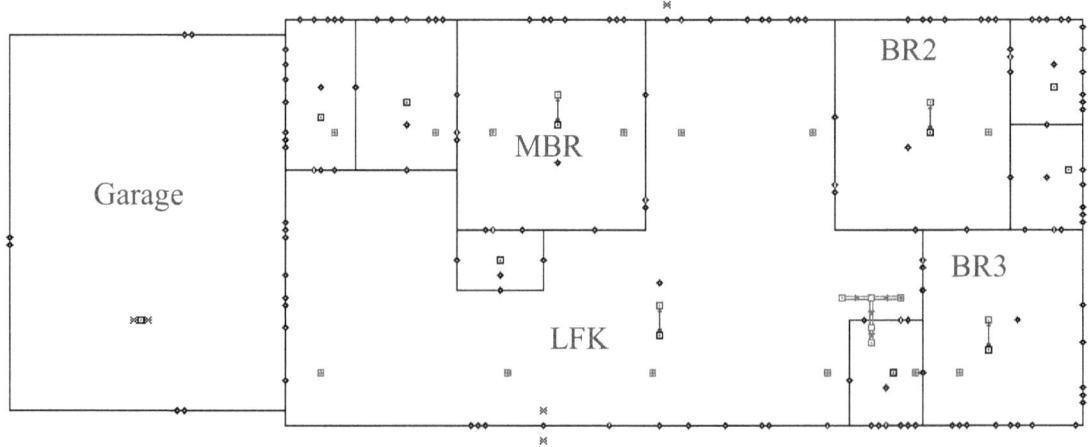

Figure 29 CONTAM Sketchpad Representation of Test House

Nabinger and Persily (2008) and Nabinger et al. (2010) describe a study of ventilation rates and energy consumption in the test house before and after a retrofit effort to improve the tightness of both the building envelope and air distribution systems. That study included fan pressurization tests of both the envelope and air distribution systems, measurements of HVAC system air flows and tracer gas decay tests of air change rates. A CONTAM airflow model of the building was created and the predicted infiltration rates agreed well with the measured values over a range of weather and system operation conditions. Tables 5 and 6 list the air leakage values (in terms of the effective leakage area (ELA) at a reference pressure difference of 4 Pa) and HVAC system airflows from Nabinger et al. (2010). The CONTAM model used a detailed duct model for the HVAC system and, as such, the modeled flow values varied somewhat from the measured values in Table 6. Also, a duct leak of about 30 L/s was included in the belly. The belly, which is located between the floor of the living space and the crawlspace, contains the supply duct work within an insulated membrane.

Table 5 Air Leakage Values

		ELA at 4 Pa
	Airflow path	Post-retrofit
Living space envelope	Exterior wall	$0.13 \text{ cm}^2/\text{m}^2$
	Ceiling wall interface	$0.73 \text{ cm}^2/\text{m}$
	Floor wall interface	$1.12 \text{ cm}^2/\text{m}$
	Window #1	5.00 cm^2
	Window #2	1.94 cm^2
	Corner interface	$0.73 \text{ cm}^2/\text{m}$
	Exterior doors	18.7 cm^2
	Living space to belly	$1.43 \text{ cm}^2/\text{m}^2$
Interior airflow paths	Interior walls	$2 \text{ cm}^2/\text{m}^2$
	Bedroom doorframe	410 cm^2
	Open interior doors	2 m x 0.9 m
	Bathroom doorframe	330 cm^2
	Interior doorframe	250 cm^2
	Closet doorframe	4.6 cm^2
Attic	Attic floor	$2 \text{ cm}^2/\text{m}^2$
	Roof vents	$0.135 \text{ m}^2/\text{each}$
	Eave vents	$296 \text{ cm}^2/\text{m}$
Crawl space and belly	Exterior walls of crawl space	$25 \text{ cm}^2/\text{m}^2$
	Rear crawl space vents	323 cm^2
	Front crawl space vents	465 cm^2
	Crawl space access door	206 cm^2
	Crawl space to "belly"	181 cm^2
	Duct leak into belly	58 cm^2

Table 6 HVAC System Supply Flows

Room	Vent No.	Airflow rate L/s Post-retrofit
Family	1	54
Family	2	44
Kitchen	3	39
Dining	4	41
Bath2	5	40
Bed3	6	62
Bed2	7	21
Living	8	27
Living	9	28
Bed1	10	29
Bed1	11	25
Bath1	12	17
Utility	13	24
Total		**450**

The CONTAM airflow model described by Nabinger et al. 2010 was modified to add the attached garage, including leakage paths from the garage to ambient and to the house. The modeled garage leakage was based on fan pressurizations tests conducted in April, July and August 2008 with analysis following the methods described by Emmerich et al. 2003.

Garage Pressurization Tests

Additional pressurization tests were conducted on the house to measure the leakage of the house, garage, and the house-garage (HG) interface. The pressurization tests were generally conducted according to ASTM Standard E 779-10 (ASTM 2010). Three configurations were used on the test house as shown in Figure 30:

1. Blower door (or pressurization fan) in the living space with the garage door open.

2. Blower door in the garage access door with the living space doors open.

3. Blower door in the living space with the HG interface door open.

In the figure and the analysis, each building is represented as two zones; a house zone (H) and a garage (G) zone separated by the house-garage interface. Arrows indicate the location of the blower door for each test and the direction of the airflow from the blower door fan. The pressure difference, ΔP_{HG}, is the pressure difference across the house-garage interface. The pressure differences across the living space exterior envelope for each test configuration are designated as ΔP_H, while ΔP_G designates the pressure differences across the garage exterior envelope.

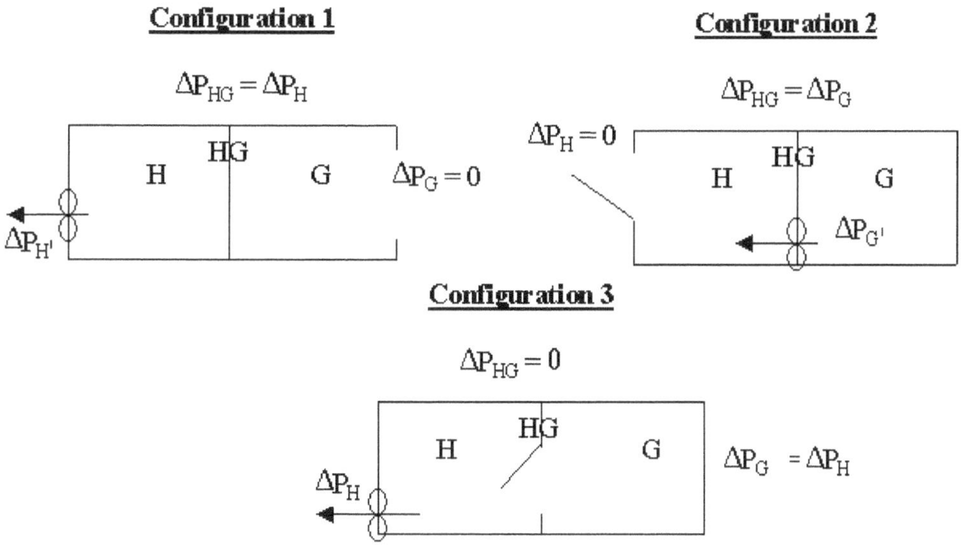

Configuration 1

$$\Delta P_{HG} = \Delta P_H$$

HG

H   G $\Delta P_G = 0$

$\Delta P_{H'}$

Configuration 2

$$\Delta P_{HG} = \Delta P_G$$

$\Delta P_H = 0$

HG

H   G

$\Delta P_{G'}$

Configuration 3

$$\Delta P_{HG} = 0$$

HG

H   G $\Delta P_G = \Delta P_H$

ΔP_H

Figure 30 Test house and garage fan pressurization configurations

---

Symbol Legend for Figures and Equations in this section

H – exterior envelope of house (living space)
H' – house exterior envelope and HG interface combined (H+HG)
G – exterior envelope of garage
G' – garage exterior envelope and HG interface combined (G+HG)
HG – HG interface
$\mathbf{Q_{\# \text{ or } L}}$ – Airflow rate from blower door in configuration # or for surface designation (L)
$\mathbf{\Delta P_{\# \text{ or } L}}$ – Pressure difference across a surface
$\mathbf{C_{\# \text{ or } L}}$ – flow coefficient for $Q_{\# \text{ or } L}$

For each blower door test, measured airflows were recorded at multiple pressures ranging from 10 Pa to 70 Pa (not all spaces were pressurized to 70 Pa). The airflows and pressures were then fit to a power law equation: $Q = C (\Delta P)^n$. For configuration 1, depressurizing the living space (H), while the garage door is open, effectively changes the garage space to an ambient zone. The house zone is bounded by the exterior envelope surfaces of the living space, H, and the HG interface surface. The reverse is performed for the garage in test 2, in which the garage zone is bounded by the exterior envelope surfaces of the garage, G, and the HG interface surface. Finally, in configuration 3, the combined house and garage spaces are pressurized.

As described in Emmerich et al. 2003, calculations for the test house are based on continuity, effective leakage area, and power law orifice equations. Applying the law of conservation of mass and the power law orifice equation yields a three by three matrix of airflow rates:

$$Q_1 = C_1 \cdot \Delta P^{n_1} = Q_H + Q_{HG} \tag{4}$$

$$Q_2 = C_2 \cdot \Delta P^{n_2} = Q_G + Q_{HG} \tag{5}$$

$$Q_3 = C_3 \cdot \Delta P^{n_3} = Q_G + Q_H \tag{6}$$

Here, C_1 and n_1 are the flow parameters obtained from the curve fit to the data for configuration 1 (tested twice in depressurization mode only), C_2 and n_2 are obtained from the data for configuration 2 (tested 3 times in depressurization mode and once in pressurization mode), and C_3 and n_3 are obtained from the data for configuration 3 (tested twice in depressurization mode only). With three equations and three unknowns, this matrix can be solved for Q_H, Q_G, and Q_{HG}, all at a specific reference pressure across surfaces H, G, and HG. Using these three airflow rates, the zone envelope and HG interface ELA's are calculated using the effective leakage area equation (ASHRAE 2009):

$$ELA = \frac{Q}{\sqrt{\left(2 \cdot \Delta P / \rho\right)}} \qquad (7)$$

Table 7 Fan Pressurization Test Results

C_1 (m^3/h·Pa^n)	290
C_2 (m^3/h·Pa^n)	157
C_3 (m^3/h·Pa^n)	296
n_1	0.61
n_2	0.56
n_3	0.65
$ACH_{50,H}$ (h^{-1})	11.8
$ACH_{50,G}$ (h^{-1})	15.5
$ELA_{4,H}$ (cm^2)	574
$ELA_{4,G}$ (cm^2)	210
$ELA_{4,HG}$ (cm^2)	157

The total leakage from the house to the garage ($ELA4,HG$) of 157 cm^2 was equal to the average total leakage from the house to the garage for the five houses tested by Emmerich et al. (2003). The effective leakage areas for G and HG from Table 7 were distributed vertically along the walls of the garage and on the ceiling in the CONTAM model.

Temperatures were measured in most of the individual rooms, garage, crawlspace and attic of the test house during the tests and used as input to the model. Temperatures in the LFK zone were based on an average of the living room, family room, and kitchen measurements. Both BR2 and BR3 temperatures were based on measured BR3 temperatures as there was no measurement available for BR2. Temperatures in the Utility zone were based on the adjacent family room temperature except when the door to the garage was open in which case they were based on an average of the temperature in the LFK zone and the garage. Temperatures in other zones without measurements such as bathrooms were based on an average of all room (i.e., not including garage, crawlspace or attic) measurements.

Model Validation

This section describes the model validation efforts that were carried out to establish the ability of the CONTAM model to predict CO levels in the garage and the house and to develop an estimate of its uncertainty of these predictions relative to the measured values. A number of past studies (e.g., Emmerich and Nabinger (2001) and Emmerich et al. (2004)) have also examined the issue of multizone IAQ model validation as reviewed in Emmerich (2001).

Statistical Evaluation of Model Predictions

As part of this validation effort, predicted CO concentrations were compared with measured values from a series of experiments in the test house. The predictions and measurements were compared using ASTM D5157 Standard Guide for Statistical Evaluation of Indoor Air Quality Models, which provides quantitative and qualitative tools for evaluation of IAQ models (ASTM 2008). It provides guidance in choosing data sets for model evaluation and focuses on evaluating the accuracy of indoor concentrations predicted by a model. As part of the comparison of CONTAM predictions to measurements of CO and O_2 concentrations, the ASTM D5157 suggested criteria were applied. The data sets collected during this study meet the ASTM D5157 criteria for model evaluation, as they are entirely independent of the data used to develop the model and to estimate model inputs. Also, the data are of sufficient detail to evaluate the CONTAM predictions of individual zonal CO concentrations.

ASTM D5157 provides three statistical tools for evaluating the accuracy of IAQ model predictions and two additional statistical tools for assessing bias. Values for these statistical criteria are provided to indicate whether the model performance is adequate. The tools for assessing agreement between predictions and measurements, and the criteria for model evaluation, include:

1) The correlation coefficient of predictions and measurements should be 0.9 or greater.

2) The line of regression between the predictions and measurements should have a slope between 0.75 and 1.25 and an intercept less than 25 % of the average measured concentration.

3) The normalized mean square error (NMSE) should be less than 0.25. The NMSE is calculated as:

$$NMSE = \sum_{i=1}^{N}(C_{pi}-C_{oi})^2 / 2\overline{C}_o\overline{C}_p \tag{8}$$

where C_p is the predicted concentration and C_o is the observed concentration, and the over-bar represents an average over the N data points during the test period for each test case.

ASTM D5157 also provides two statistical measures of bias with values for judging adequate model performance. These measures of bias include:

1) Normalized fractional bias (FB) of the mean concentrations. Fractional bias should be 0.25 or lower and is calculated as:

$$FB = 2(\overline{C}_p - \overline{C}_o)/(\overline{C}_p + \overline{C}_o) \tag{9}$$

2) Fractional bias based on the variance (FS) which should be 0.5 or lower. FS is calculated as:

$$FS = 2(\sigma_p{}^2 - \sigma_o{}^2)/(\sigma_p{}^2 + \sigma_o{}^2)$$ (10)

where σ_p is the standard deviation of the predicted concentrations and σ_o is the standard deviation of the observed concentrations.

Testing Configurations

As previously discussed in the Garage Test Results section and as part of the model validation effort, testing was conducted under seven different test house configurations to evaluate their impacts on the ability to predict the build-up of CO in the garage and its transport into the different rooms in the house. These configurations included two different garage bay door positions (fully closed or open nominally 0.6 m), two connecting door positions between the garage and the house (fully closed or open nominally 5 cm), and two house central heating, ventilating, and air conditioning (HVAC) fan settings (on or off). When the garage bay door was open, there was also a vertical opening of approximately 0.15 m at the top of the bay door. All internal house doors were kept open throughout all validation tests and model analyses.

Table 8 includes a summary of the validation tests conducted including information on the generator tested, the test house configuration (door positions and fan status), a test identification code, the date the test was conducted, the average ambient temperature and wind speed, and the CO analyzers used. The validation tests in Table 8 with Gen SO1 were included in the garage test results section, but those with Gen B were used exclusively for model validation.

Table 8 Model Validation Tests Conducted

Generator	Garage bay door	Garage to house entry door	HVAC fan	Date	Outdoor Temp (°C)	Wind speed (m/s)	Load	CO analyzers in garage	CO analyzers in house
B	Closed	Open	OFF	10/8/08	18.4	4.5	2.5 kW	N1, T1	N2, N3
B	Open	Open	OFF	10/23/08 a.m.	8.8	2.2	5 kW	N1, T1	N2, N3
B	Closed	Closed	ON	10/23/08 p.m.	13.1	3.8	5 kW	N1, T1	N2, N3
SO1 cat	Closed	Open	OFF	04/01/10	19.9	6.3	Cyclic	N2, N3	N1, R1
SO1 cat	Open	Open	ON	04/22/10	20.4	7.8	Cyclic	N2, N3	N1, R1
SO1 cat	Closed	Closed	ON	04/29/10	17.8	9.5	Cyclic	N2, N3	N1, R1
SO1 with noncat muffler	Closed	Closed	OFF	05/13/10	15.6	6.5	Cyclic	N2, N3	N1, R1

RESULTS OF MODEL VALIDATION

As shown in Table 8, seven validation cases were simulated under a variety of conditions between October 2008 and May 2010. The tests consisted of operating one of the generators in the attached garage and measuring the concentrations in the house and garage for two to six hours.

Case #1

In the first validation case, Gen B was operated with a nominal load of 2.5 kW for 2 h in the garage with the garage bay door closed, house entry door open and the HVAC system off. Based on the shed test results and the fact that the O_2 in the garage did not drop below about 19 % during this test, a constant CO emission rate of 760 g/h and a constant O_2 consumption rate of 4140 g/h were applied in the simulations of this case. These values are based on the average shed test results plotted on Figures 8a and 8b for O_2 at 19 % or above.

The observed and predicted CO and O_2 concentrations for the garage zone and the LFK zone in the house are shown in Figure 31. Table 9 shows average observed and predicted zone CO concentrations and percent differences, along with the ASTM D5157 statistical parameters (described previously) calculated for the zone average concentrations. The suggested ASTM D5157 statistical criteria were evaluated for overall zone average concentrations for the entire testing period for all cases. The columns of Table 9 (and subsequent tables) include the zone average observed concentration (C_o), average predicted concentration (C_p), standard deviation of observed concentrations (σ_o), standard deviation of predicted concentrations (σ_p), correlation coefficient (R), regression slope (m), regression intercept divided by the average observed concentration (b/C_o), normalized mean square error (NMSE), fractional bias of the mean concentrations (FB), and fractional bias based on the variance (FS). The average concentration in the bottom row of the tables is a linear average of the five zone concentrations.

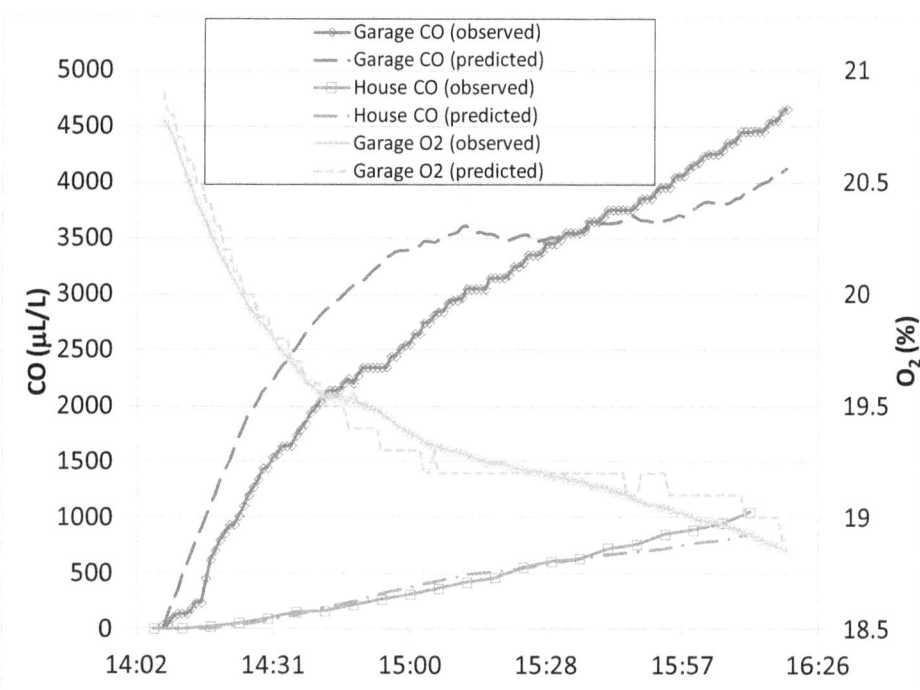

Figure 31 Predicted and observed CO and O_2 concentrations for Case #1 (Notes: Plotted house concentration is from the LFK zone for Figures 31 through 37. The x-axis shows clock time for Figures 31 through 37.)

Table 9 Statistical parameters for Case #1 (concentrations in units of $\mu L/L$)

Zone	C_o	C_p	C_o-C_p	% diff							
GAR	2802	3066	-265	9							
LFK	435	414	21	-5							
MBR	307	337	-30	10							
Bed2	308	383	-75	24							
Bed3	289	369	-81	28							
Average |% difference|				**15**							
	C_o	C_p		σ_o	σ_p	R	m	b/C_o (%)	NMSE	FB	FS
Room Average Concentrations	828	914		1100	1200	**1.00**	**1.09**	**1.5**	**0.02**	**0.10**	**0.27**

The bold values in Table 9 are those that meet the ASTM suggested criteria. Based on the statistical parameters, this case exhibits excellent agreement between measurements and predictions. Specifically, the values for R, m, b/Co, NMSE, FS and FB calculated for the average zone concentrations all meet the ASTM D5157 suggested limits. Additionally,

65

the average of the absolute percent differences between the zone values of C_o and C_p was 15 %.

Case # 2

Case # 2 involved operation of Gen B in the garage with the bay door open, the house entry door open and the HVAC system fan off. A constant nominal load of 5 kW was applied for this case. Based on the shed test results and the fact that the O_2 in the garage did not drop below 20.8 % during this test, a constant CO emission rate of 1090 g/h (corresponding to the highest O_2 point during the shed testing [see Fig 6a]) and a constant O_2 consumption rate of 5640 g/h (see Fig 6b) were applied to the simulations.

The observed and predicted concentrations for garage zone and the LFK zone in the house are shown in Figure 32, and the calculated D5157 statistical parameters are shown in Table 10. Based on the statistical parameters, the agreement between measurements and predictions was not as good as for Case #1. Specifically, only the values for R and FB calculated for the average zone concentrations meet the ASTM D5157 suggested limits while m, b/Co, NMSE, and FS fall outside the limits. Also, the average of the absolute percent differences between the zone values of C_o and C_p, i.e., 24 %, was worse than Case 1. This larger average value was driven primarily by a large difference in the garage zone, though smaller differences are seen in the house zones. One possible reason for poorer agreement is the lack of shed test data for an O_2 level above 20.5 % (see Figure 8a).

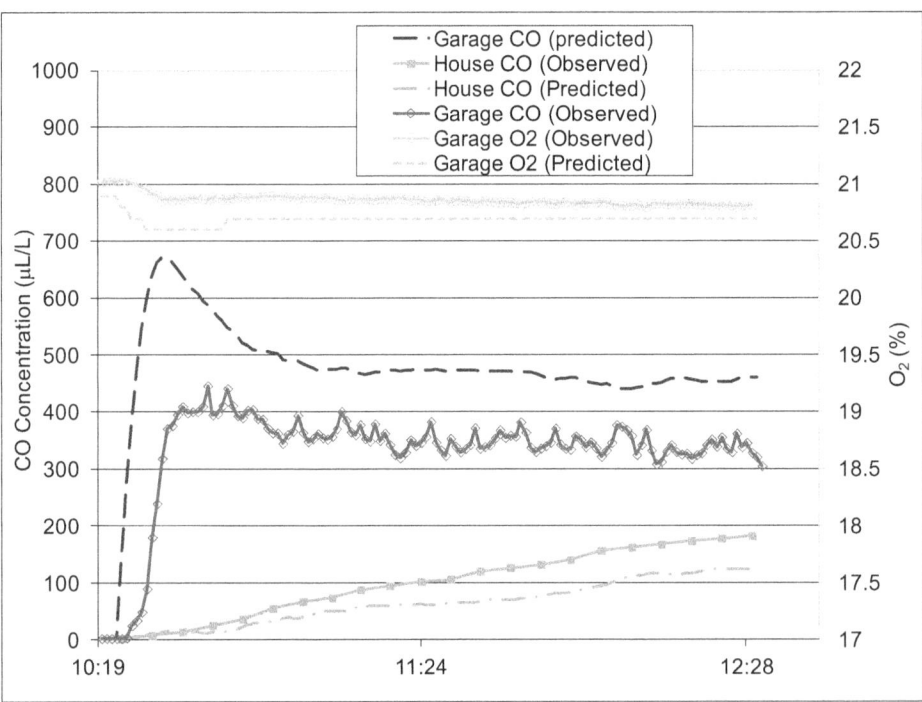

Figure 32 Predicted and observed CO and O_2 concentrations for Case # 2

Table 10 Statistical parameters for Case # 2 (concentrations in units of μL/L)

Zone	C_o	C_p	C_o-C_p	% diff								
GAR	329	470	-142	43								
LFK	96	61	34	-36								
MBR	55	49	6	-11								
Bed2	73	52	21	-28								
Bed3	50	51	-1	2								
Average |% difference|				24								
	C_o	C_p		σ_o	σ_p	R	m	b/C_o (%)	NMSE	FB	FS
Room Average Concentrations	120	137		118	186	**0.99**	1.57	-43	0.26	**0.19**	1.0

Case # 3

Case # 3 involved operation of Gen B in the garage with the bay door closed, the house entry door closed and the HVAC system fan on. A constant nominal load of 5 kW was applied for this case. Based on average values from the shed test results in Figure 8, a constant O_2 consumption rate of 5940 g/h was applied and the CO emission rate was dependent on the predicted O_2 level per the relationship in Table 11.

Table 11 O_2 dependent CO emission rate for Gen B

O_2 level (%)	CO emission (g/h)
O_2>20	1370
19<O_2<20	2070
18<O_2<19	3000
17<O_2<18	3150
O_2<17	2140

The observed and predicted transient concentrations for the garage zone and the LFK zone in the house are shown in Figure 33, and the D5157 statistical parameters calculated for the zone average concentrations are shown in Table 12. Based on the statistical parameters, this case resulted in agreement between measurements and predictions similar to that seen in Case 2. Specifically, the values for R and b/Co calculated for the comparison of average zone concentrations meet the ASTM D5157 suggested limits while those for m, NMSE, FB, and FS fall outside the limits. However, the average of the absolute percent differences between C_o and C_p, i.e., 16 %, was better than Case 2. Again, there are large differences in the garage zone concentrations, but differences in the house zones are smaller.

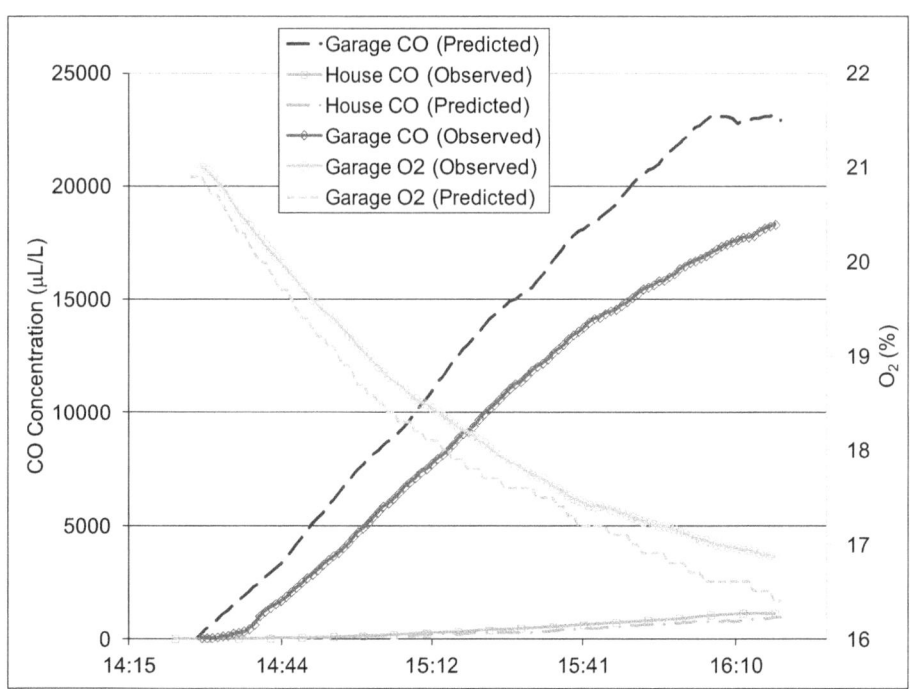

Figure 33 Predicted and observed CO and O_2 concentrations for Case # 3

Table 12 Statistical parameters for Case # 3 (concentrations in units of $\mu L/L$)

Zone	C_o	C_p	C_o-C_p	% diff							
GAR	9542	12978	-3436	36							
LFK	434	322	111	-26							
MBR	275	254	20	-7							
Bed2	289	289	0	0							
Bed3	262	289	-27	10							
Average |% difference|				16							
	C_o	C_p		σ_o	σ_p	R	m	b/C_o (%)	NMSE	FB	FS
Room Average Concentration	2160	2827		4130	5580	**1.00**	1.37	**-6.6**	0.39	0.27	0.85

Case # 4

Case # 4 involved operation of Gen SO1 with the CAT muffler in the garage with the bay door closed, the house entry door open, the HVAC system fan off and a cyclic electrical load. During the test, a load bank fuse failed and the load dropped in half for a few minutes before the fuse was replaced and the test continued. Also, this test includes a natural decay period of about 40 min (included in plots and analysis below) following generator shut-off.

Based on average values from the shed test results shown in Figure 11, load-dependent O_2 consumption and CO emission rates were used for this case (see Table 13), with both rates dependent on the electrical load. In addition, past measurement of Gen SO1 operation in both the shed and garage showed that an initial spike of CO is consistently observed immediately following a cold start (note the ECU algorithm is programmed to operate rich upon start-up). Based on analysis of cold start-up data following the method described previously for shed testing, a 430 g/h source of CO was included in the model at start-up in place of the Table 13 value until the measured AFR exceeded 13.5 (the first five minutes of operation for this case).

Table 13 Hourly cyclic load profile for Gen SO1 cat

Load bank setting (W)	Duration (min)	CO emission (g/h)	O_2 consumption (g/h)
no load	3	20	2620
500	4	15	3880
1500	18	17	3080
3000	17.5	47	4820
4500	12	96	4370
5500	5.5	102	3710

The observed and predicted transient concentrations for the garage zone and LFK zone in the house are shown in Figure 34, and the D5157 statistical parameters calculated for the zone average concentrations are shown in Table 14. Based on the statistical parameters, this case resulted in agreement between average concentration measurements and predictions similar to Case #2. Specifically, only the value for R and FB calculated for the comparison of average zone concentrations met the ASTM D5157 suggested limits while m, b/Co, NMSE, and FS fall outside the limits. Additionally, the average of the absolute percent differences between zone values of C_o and C_p was 26 %, which was driven by a large difference in the garage zone but much smaller differences in the LFK zone.

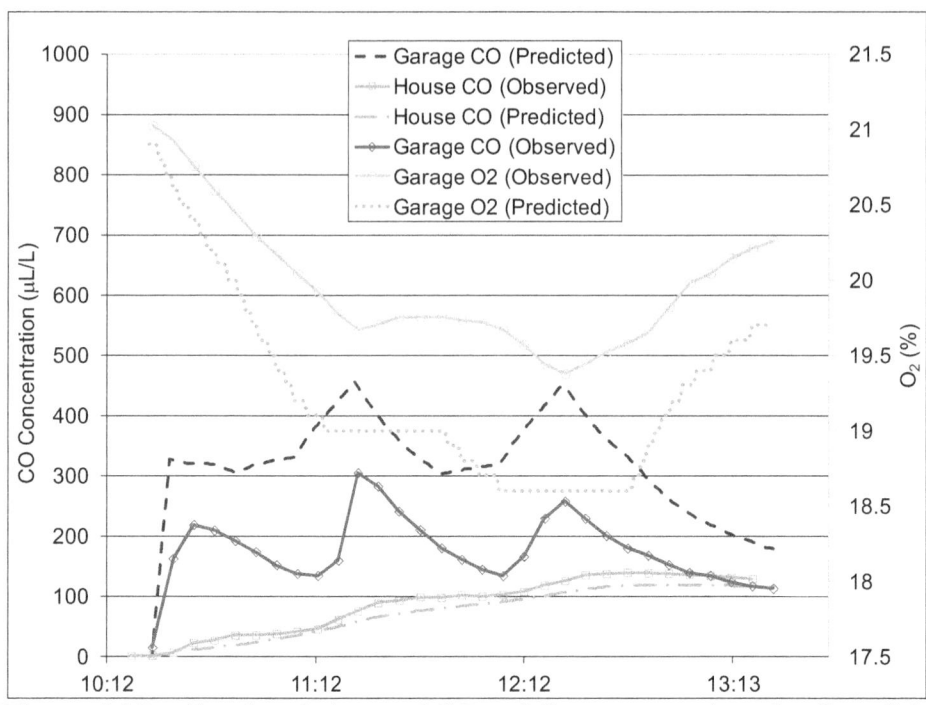

Figure 34 Predicted and observed CO and O$_2$ concentrations for Case # 4

Table 14 Statistical parameters for Case # 4 (concentrations in units of μL/L)

Zone	C$_o$	C$_p$	C$_o$-C$_p$	% diff								
GAR	175	326	-151	86								
LFK	85	73	12	-14								
MBR	83	69	14	-17								
Bed2	77	69	7	-10								
Bed3	69	70	0	1								
Average |% difference|				**26**								
	C$_o$	C$_p$		σ$_o$	σ$_p$	R	m	b/C$_o$ (%)	NMSE	FB	FS
Room Average Concentrations	98	121		43	114	**0.99**	2.6	-135	0.39	**0.21**	1.6

Case # 5

Case # 5 involved operation of Gen SO1 with CAT muffler in the garage with the bay door open, the house entry door open and the HVAC system fan on. A cyclic load was applied for this case. Also, this test includes a natural decay period of about 25 min following generator shut-off. Based on average values from the shed test results shown in Figure 11, load-dependent O$_2$ consumption and CO emission rates were used for this case (see Table 15) along with a 430 g/h source of CO at start-up in place of the Table 15 emission until the measured AFR exceeded 13.5 (ten minutes for this case).

Table 15 Hourly cyclic load profile for Gen SO1 cat

Load bank setting (W)	Duration (min)	CO emission (g/h)	O_2 consumption (g/h)
no load	3	20	2620
500	4	15	3880
1500	18	17	3080
3000	17.5	47	4820
4500	12	96	4370
5500	5.5	102	3710

The observed and predicted transient concentrations for the garage zone and the LFK zone in the house are shown in Figure 35, and the D5157 statistical parameters calculated for the zone average concentrations are shown in Table 16. Based on the statistical parameters, this case resulted in good agreement between measurements and predictions. The average absolute value of the percent differences between the zone measurements and predictions was 20 %, and the absolute difference was less than 10 µL/L for all zones. Additionally, the values for R, NMSE, and FB calculated for the comparison of average zone concentrations meet the ASTM D5157 suggested limits, while m, b/Co, and FS fall somewhat outside the limits.

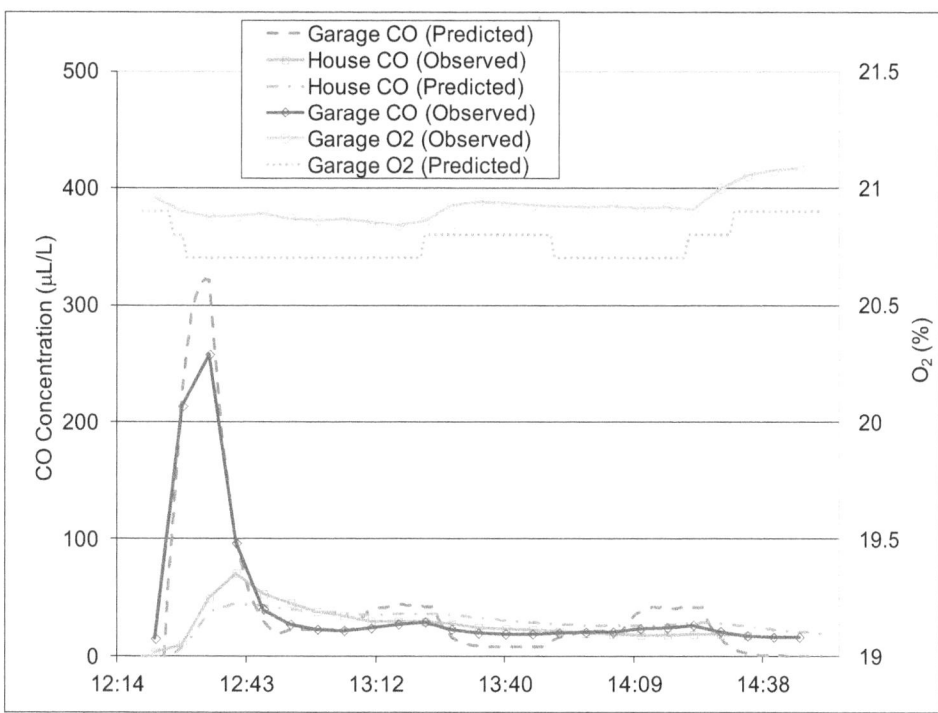

Figure 35 Predicted and observed CO and O_2 concentrations for Case # 5

Table 16 Statistical parameters for Case # 5 (concentrations in units of μL/L)

Zone	C_o	C_p	C_o-C_p	% diff						
GAR	41	41.9	-0.5	1						
LFK	26	29.2	-2.8	11						
MBR	23	28.4	-5.8	25						
Bed2	21	28.5	-7.5	36						
Bed3	23	28.7	-5.9	26						
Average |% difference|				20						
	C_o	C_p		σ_o	σ_p	R	m	b/C_o (%)	NMSE	FB	FS
Room Average Concentrations	27	31		8.4	5.9	**0.98**	0.69	48	**0.03**	**0.15**	-0.54

Case # 6

Case # 6 involved operation of Gen SO1 with CAT muffler in the garage with the bay door closed, the house entry door closed and the HVAC system fan on. A cyclic load was applied for this case. Based on the average values from the shed test results shown in Figure 11, load-dependent O_2 consumption and CO emission rates were used for this case (see Table 17) along with a 430 g/h source of CO at start-up in place of the Table 17 emission until the measured AFR reached 13.5 (ten minutes for this case). There was a problem with the Family room temperature measurement for this case so the utility room temperature was set equal to the kitchen temperature.

Table 17 Hourly cyclic load profile for Gen SO1 cat

Load bank setting (W)	Duration (min)	CO emission (g/h)	O_2 consumption (g/h)
no load	3	20	2620
500	4	15	3880
1500	18	17	3080
3000	17.5	47	4820
4500	12	96	4370
5500	5.5	102	3710

The observed and predicted transient concentrations for the garage zone and the LFK zone in the house are shown in Figure 36, and the D5157 statistical parameters calculated for the zone average concentrations are shown in Table 18. The average absolute value of the percent differences between the zone measurements and predictions was 16 %. Based on the statistical parameters, this case resulted in excellent agreement between measurements and predictions. Specifically, the values for R, m, b/Co, NMSE, FB and

FS calculated for the comparison of average zone concentrations all meet the ASTM D5157 suggested limits.

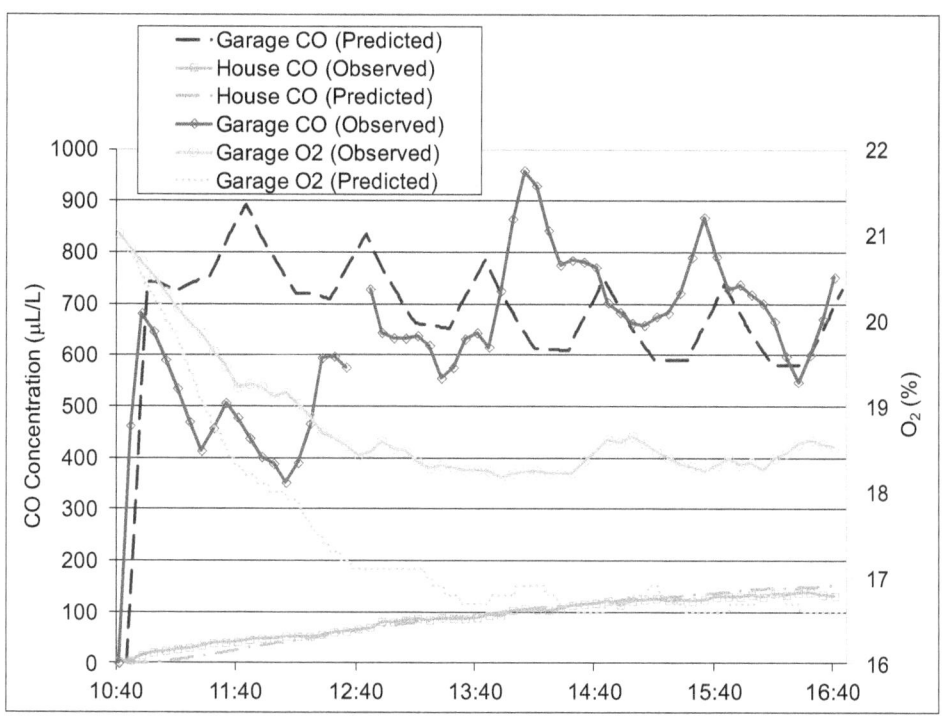

Figure 36 Predicted and observed CO and O_2 concentrations for Case # 6

Table 18 Statistical parameters for Case # 6 (concentrations in units of μL/L)

Zone	C_o	C_p	C_o-C_p	% diff							
GAR	631	686	-55	9							
LFK	87	88	0	0							
MBR	72	84	-12	17							
Bed2	63	84	-22	34							
Bed3	72	86	-13	19							
Average |% difference|				**16**							
	C_o	C_p		σ_o	σ_p	**R**	**m**	**b/C_o (%)**	**NMSE**	**FB**	**FS**
Room Average Concentrations	185	206		250	269	**1.00**	**1.07**	**3.5**	**0.02**	**0.10**	**0.25**

73

Case # 7

Case # 7 involved operation of Gen SO1 with nonCAT muffler in the garage with the bay door closed, the house entry door closed and the HVAC system fan off. A cyclic load was applied for this case. This case included a natural decay for about 45 min after generator shut-off (included in plots and analysis below). Based on average values of the shed test results shown in Figure 12, load-dependent O_2 consumption and CO emission rates were used for this case (see Table 19) along with an 840 g/h source of CO at start-up in place of the Table 19 emission until the AFR reached 13.5 (the first ten minutes of operation for this case).

Table 19 Hourly cyclic load profile for Gen SO1 noncat

Load bank setting (W)	Duration (min)	CO emission (g/h)	O_2 consumption (g/h)
no load	3	150	2580
500	4	88	2780
1500	18	97	3450
3000	17.5	236	4110
4500	12	271	4190
5500	5.5	236	4440

The observed and predicted transient concentrations for the garage zone and the LFK zone in the house are shown in Figure 37, and the D5157 statistical parameters calculated for the zone average concentrations are shown in Table 20. The average absolute value of the percent differences between the zone measurements and predictions was 17 %. Based on the statistical parameters, this case resulted in excellent agreement between measurements and predictions. Specifically, the values for R, m, b/Co, NMSE, FB and FS calculated for the comparison of average zone concentrations all meet the ASTM D5157 suggested limits.

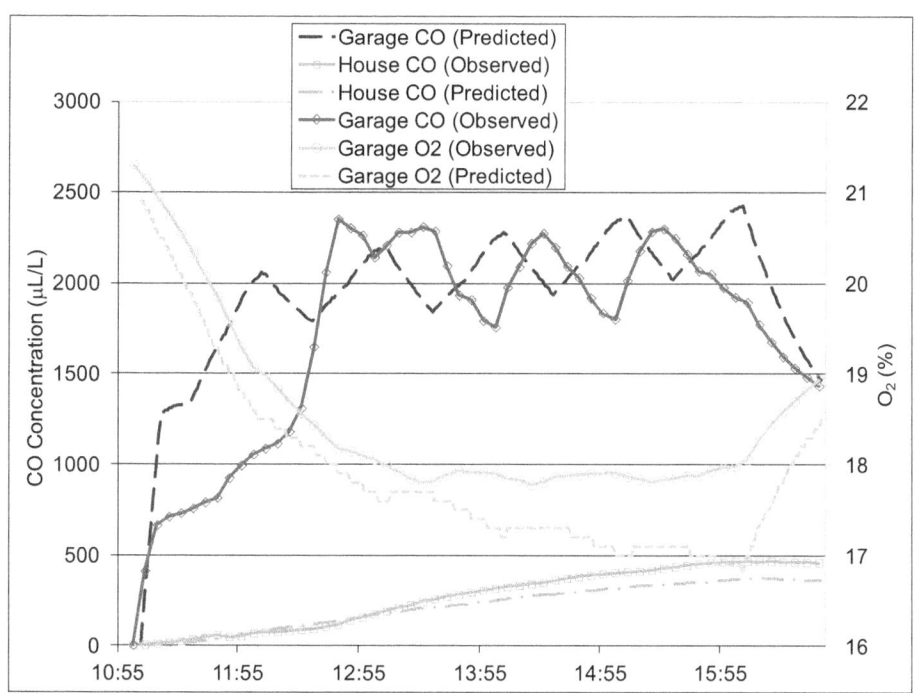

Figure 37 Predicted and observed CO and O_2 concentrations for Case # 7

Table 20 Statistical parameters for Case # 7 (concentrations in units of μL/L)

Zone	C_o	C_p	C_o-C_p	% diff							
GAR	1712	1944	-232	14							
LFK	266	223	42	-16							
MBR	264	212	52	-20							
Bed2	263	216	47	-18							
Bed3	261	216	46	-18							
Average |% difference|				17							
	C_o	C_p		σ_o	σ_p	R	m	b/C_o (%)	NMSE	FB	FS
Room Average Concentrations	553	562		648	772	1.00	1.19	-18	0.04	0.02	0.36

Summary

Table 21 summarizes the statistical parameters calculated for all the cases based on the average zone concentrations. The bold values in Table 21 are those that meet the ASTM suggested criteria. As seen in the table, statistical agreement varies for the individual cases. The agreement was generally better for the cases with both the bay and house entry doors closed (3, 6, and 7). This improved agreement is not a surprise, as the cases with better agreement rely primarily on CONTAM's ability to model airflow through smaller leaks. While CONTAM includes elements to model flow through larger openings (used for the open bay and entry doors of Cases 1, 2, 4 and 5), those flow predictions are thought to be less reliable given the physical theory on which CONTAM is based. Table 21 shows that the statistical agreement was generally somewhat worse for those cases.

Table 21 Summary of statistical parameters from comparison of average observed and predicted zone CO concentrations for all cases

Case	C_o	C_p	Avg |%diff|	σ_o	σ_p	R	m	b/C_o (%)	NMSE	FB	FS
1	828	914	15	1100	1200	**1.00**	**1.09**	**1.5**	**0.02**	**0.10**	**0.27**
2	120	137	24	118	186	**0.99**	1.57	-43	0.26	**0.19**	1.0
3	2160	2827	16	4130	5580	**1.00**	1.37	**-6.6**	0.39	0.27	0.85
4	98	121	26	43	114	**0.99**	2.6	-135	0.39	**0.21**	1.6
5	27	31	20	8.4	5.9	**0.98**	0.69	48	**0.03**	**0.15**	-
6	185	206	16	250	269	**1.00**	**1.07**	**3.5**	**0.02**	**0.10**	**0.25**
7	553	562	17	648	772	**1.00**	**1.19**	-18	**0.04**	**0.02**	**0.36**
Overall summary statistics for all tests											
Garage CO	2180	2790	28	3400	4620	**1.00**	1.36	**-7.6**	0.28	**0.25**	0.81
Garage O_2	19.5	19.1	2.5	1.1	1.3	**0.94**	**1.18**	-20	**0.001**	**-0.02**	**0.41**
House zones CO	165	160	17	130	128	**0.96**	**0.94**	**3.0**	**0.05**	**-0.03**	**-0.08**

While the statistical evaluation of the individual cases is useful to understand strengths and weaknesses of the model, a more important statistical evaluation is the comparison of the entire set of cases. This overall evaluation indicates the model's ability to predict the relative outcome when individual parameters are changed (e.g., door open vs. closed, or one type of generator vs. another). The last three rows of Table 21 include summary statistics for the average CO and O_2 concentrations in the garage and for the average CO calculated for the house zones. The agreement between the measurements and predictions for this entire set of cases was excellent for both the garage O_2 concentrations and the house CO concentrations. All of these statistical values met the D5157 criteria. The agreement, however, was somewhat worse for the garage CO concentrations, with some parameters falling slightly outside the ASTM criteria limits. The average concentrations for individual house zones (i.e., LFK, MBR, BR2 and BR3) from all cases are plotted in Figure 38 to show the comparison of predictions and measurements. To summarize, average individual house zone and garage CO concentration predictions and measurements were within about 20 % and 30 % respectively when averaged over all cases.

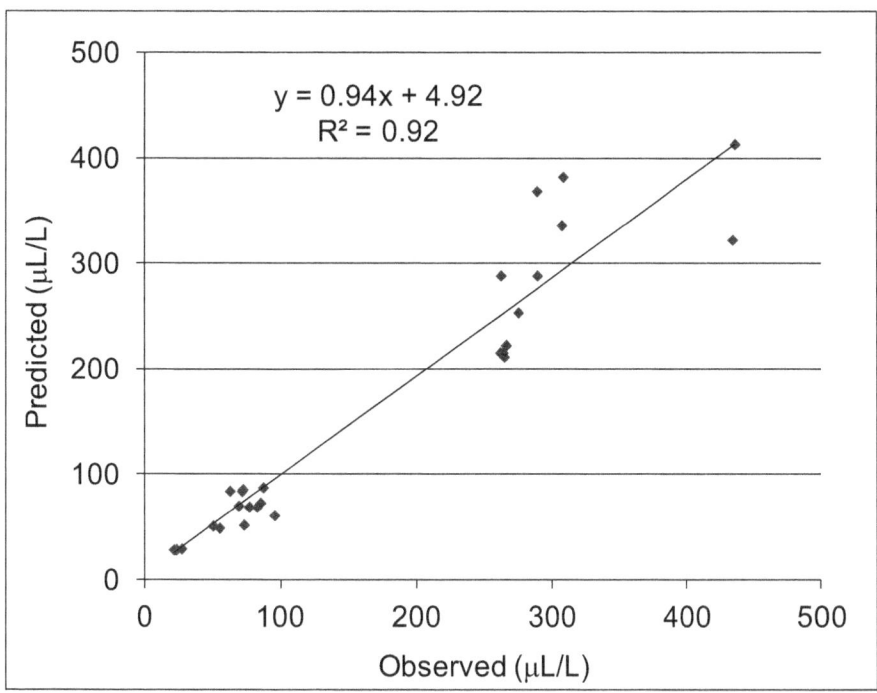

Figure 38 Comparison of predicted and observed individual house zone average CO concentrations for all cases

As discussed previously (Emmerich 2001), absolute validation of a complex model, such as CONTAM, is impossible as there are infinite possible building models that can be created by a user. However, the primary objective of this experimental validation is to evaluate the accuracy of the CONTAM model predictions for this specific application, to identify large sources of error, and to determine the level of confidence that we may have in the predictions for the other cases simulated. For the cases modeled in this effort, no significant errors in the CONTAM model were identified. Additionally, some of the discrepancies between model predictions and experimental measurements may be due to measurement limitations instead of model deficiencies. For example, this effort involved a fairly rich data set in terms of number of variables monitored and spatial and temporal detail. Still, there are additional measurements that would have been helpful for the simulations. Specifically, the utility room provided the primary pathway for contaminant transport from the garage into the house; however, temperatures and concentrations were not measured in this zone. Also, uncertainties in experimental measurements include more than simply the instrument accuracy. Many of the reported measurements are a single point that was used to represent an average room concentration. The ability of this single point measurement to represent the room may be questionable at times.

Prototype Generator Performance Simulations

As previously mentioned, a primary objective of these simulations was to examine the potential performance of the prototype generator under a wider range of conditions than studied during the experiments at the test house, such as a longer period of generator operation and operation in a house location other than the garage. These additional simulations were performed to allow analyses of the safety implications of operating a portable generator in these conditions. All of these simulations, a total of 42, were based on the NIST manufactured test house model and the tested generators, which were extensively validated as described above.

Model parameters that were varied included ambient conditions, CO emission rates, source locations (location of the operating generator) and door positions. All simulations used steady state weather with three ambient temperatures (mild, 15 °C; cold, 0 °C; and hot 30 °C) and one wind condition (2 m/s blowing towards the garage bay door).

Since the garage measurements reported above clearly establish the danger of operating the unmodified generators in an attached garage, it was decided to focus the simulations on the potential performance of the reduced CO emission prototype. Therefore, the CO emission rates in the simulation were based on operation of Gen SO1 (both with and without cat muffler) under the cyclic load profile for 18 hours followed by 6 hours with the generator off. The 18 hour period of operation is a conservative estimate based on the fuel consumption rate at a low load and a 5 gallon gas tank capacity; the actual run time would likely be shorter under real operation but using 18 hours provides a more conservative estimate of the CO levels and exposure. Based on the shed test results, the previously-validated load-dependent CO emission rates of Table 13 were used for Gen SO1 with cat muffler along with a constant 430 g/h of CO at start-up in place of the Table 13 emission for the first 10 minutes of operation (typical time to reach an AFR of 13.5 for the model validation cases). Similarly, the previously-validated, load-dependent CO emission rates of Table 19 were used for Gen SO1 with noncat muffler along with a constant 840 g/h of CO at start-up for the first 10 minutes of operation.

The simulations included the generator located in either the attached garage or in the utility room of a modified test house model with the garage removed. For garage source cases, 4 door position combinations were considered: bay door (closed or 0.6 m open) and garage/utility access door positions (closed or 5 cm open). For the utility source cases, two positions of the utility/family room connecting door were modeled (closed and 5 cm open). The temperature, source strength, location and door factors resulted in a matrix of 36 cases. See Tables 22 and 23 for lists of the garage and utility source cases, respectively.

The HVAC system fan was modeled as off for all cases. Typically all other interior house doors were modeled as open but, for a set of 6 additional cases, the bedroom doors were modeled as closed (See Table 24 for closed bedroom door cases). Interior zone temperatures were set at 23 °C except for source zones and zones adjacent to source zones. Source zones were initially 23 °C, then experienced a linear increase over 2 h to 40 °C, which was held constant until the generator stopped operating, followed by a linear decrease over the final 6 h back to 23 °C. Zones adjacent to the source zone experienced a linear increase over 2 h to 30 °C until the generator stops operating and a linear decrease over the final 6 h to 23 °C.

Garage source cases

Figure 39 presents the simulation results for the cold, mild and hot ambient temperature cases with Gen SO1 with catmuffler in the garage with the bay door open and the house door closed. For these cases, the CO concentration in the garage spikes to around 200 µL/L to 300 µL/L, but then drops and remains well below 100 µL/L and does not exceed 30 µL/L in the house. It can also be seen in Figure 39 that the temperature difference between the rooms of the house with the open doors resulted in very uniform concentrations regardless of ambient temperature. The other garage source cases have mostly similar patterns with the primary difference among all the cases with different door positions being the peak garage and house concentrations reached.

Table 22 presents the peak CO concentrations in the house (not including utility room) and garage for all of the garage source cases. In general, for a specific set of door positions, increasing the ambient temperature from cold to mild to hot resulted in higher garage and house CO concentrations with the impact of a change from cold to hot being as large as a factor of 3. The results for cases 4 through 6 show that adding the open house door increased the peak house CO concentration, but the concentrations were still at or below 50 µL/L in the house.

As expected, the results for cases 7 through 9 with the garage source, closed bay door, closed house door and cat muffler show that with the garage bay door closed the CO concentration reaches much higher peak concentrations in both the garage (850 µL/L to 1300 µL/L) and house (30 µL/L to 100 µL/L) compared to the garage bay door open cases. As shown for cases 10 through 12, opening the house door results in slightly lower garage CO (710 µL/L to 810 µL/L) but higher house CO (70 µL/L to 160 µL/L) compared to cases 7 through 9 as more CO migrates to the house from the garage through the open door.

The larger CO source for the prototype with noncat muffler (cases 13 through 24 in Table 22) resulted in peak garage CO concentrations approximately 2 to 3 times higher than for the equivalent case with cat muffler. The increase in peak house CO concentrations due to the noncat muffler source ranged from a factor of 3 to 4. The noncat muffler source results also display the same impacts due to door opening/closing and ambient temperature as the cat muffler source cases. The highest peak garage CO concentrations were for the noncat source with closed bay door and closed house door at 4300 µL/L while the highest peak house CO concentration was for the noncat source with closed bay door and open house door at 530 µL/L.

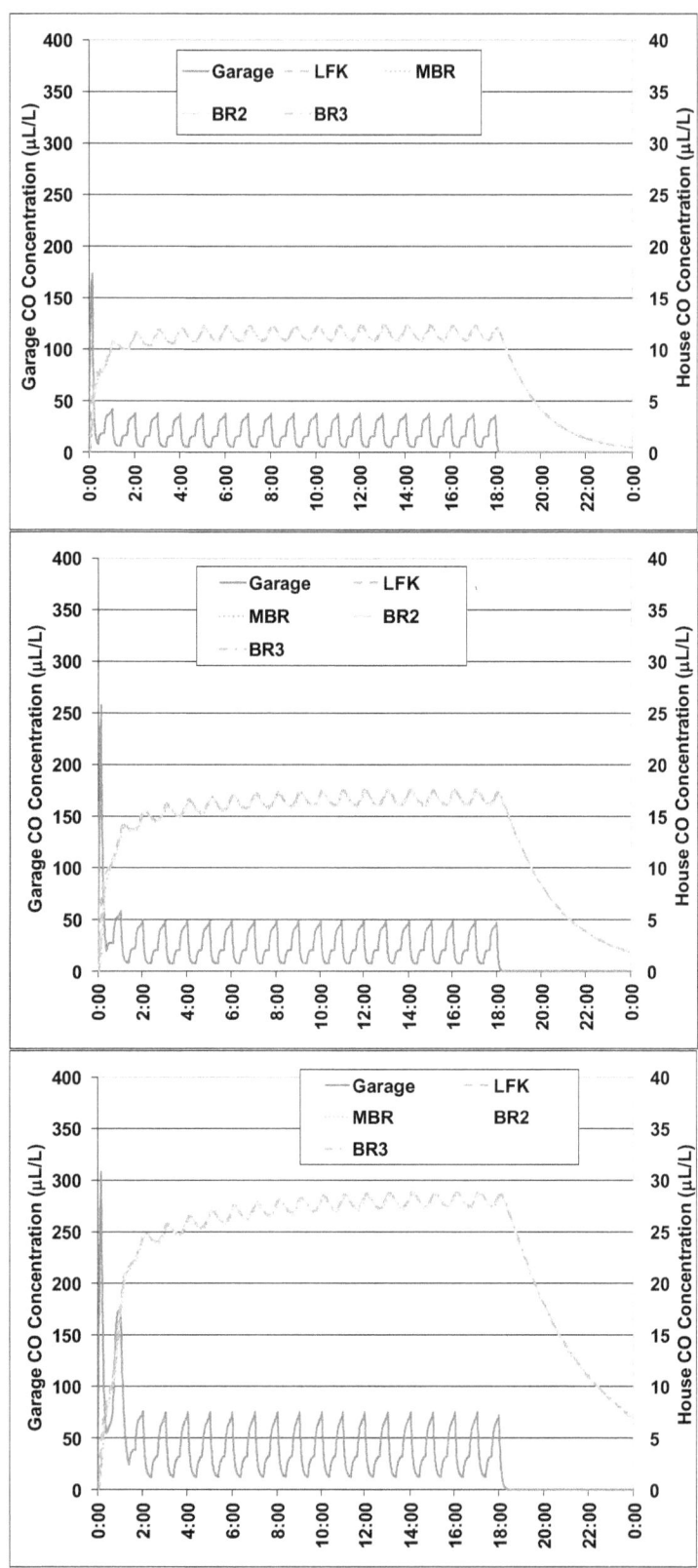

Figure 39 Garage source, open bay door, closed house door, cat muffler cases (a) cold (0 °C), (b) mild (15 °C), (c) hot (30 °C) ambient temperature

Table 22 Simulation results for garage source cases

Sim ID	Source location	Generator	Bay door	House Door	Weather	Garage Peak CO Concentration (μL/L)	House Peak CO Concentration, (Excluding Utility Room) (μL/L)
1	garage	SO1 cat	open	closed	cold	170	10
2	garage	SO1 cat	open	closed	mild	260	20
3	garage	SO1 cat	open	closed	hot	300	30
4	garage	SO1 cat	open	open	cold	170	20
5	garage	SO1 cat	open	open	mild	240	30
6	garage	SO1 cat	open	open	hot	300	50
7	garage	SO1 cat	closed	closed	cold	850	30
8	garage	SO1 cat	closed	closed	mild	1000	60
9	garage	SO1 cat	closed	closed	hot	1300	100
10	garage	SO1 cat	closed	open	cold	710	70
11	garage	SO1 cat	closed	open	mild	770	110
12	garage	SO1 cat	closed	open	hot	810	160
13	garage	SO1 noncat	open	closed	cold	330	40
14	garage	SO1 noncat	open	closed	mild	500	60
15	garage	SO1 noncat	open	closed	hot	600	100
16	garage	SO1 noncat	open	open	cold	310	70
17	garage	SO1 noncat	open	open	mild	470	90
18	garage	SO1 noncat	open	open	hot	600	150
19	garage	SO1 noncat	closed	closed	cold	2700	100
20	garage	SO1 noncat	closed	closed	mild	3300	190
21	garage	SO1 noncat	closed	closed	hot	4300	350
22	garage	SO1 noncat	closed	open	cold	2000	250
23	garage	SO1 noncat	closed	open	mild	2300	380
24	garage	SO1 noncat	closed	open	hot	2600	530

Utility source

Simulations were also performed with the generator operating in the living space of the house as opposed to the garage. Figure 40 presents the simulation results for the cold, mild and hot ambient temperature cases with Gen SO1 operating in the utility room with the door closed. For these cases, the CO concentration in the house (not including the utility room in which the peak concentration exceeds 5000 μL/L) reaches a peak of 280 μL/L to 520 μL/L. The variation on the house peak CO concentration due to variations in the ambient temperature was close to a factor of two, with higher concentration associated with higher temperature. The other utility room source cases (listed in Table 23) showed a similar effect of ambient temperature. However, there was no notable effect on the CO distribution among the different rooms in the house.

Table 23 presents the peak CO concentrations in the house (not including utility room) and in the utility room (not including the initial concentration spike) for all of the utility source cases. The results for cases 28 through 30 show that adding the open utility door increased the peak house CO concentrations by up to 20 % compared to the closed utility door cases. As seen in Table 23 for cases 31 through 36, the larger CO source for the prototype with noncat muffler results in the peak CO concentration in the house (not including the utility room) increasing by a factor of 3 to 4 times the equivalent cat muffler cases. The worst case is the prototype with noncat muffler in the utility room with the door open and hot weather which results in a peak house CO concentration of about 2000 μL/L.

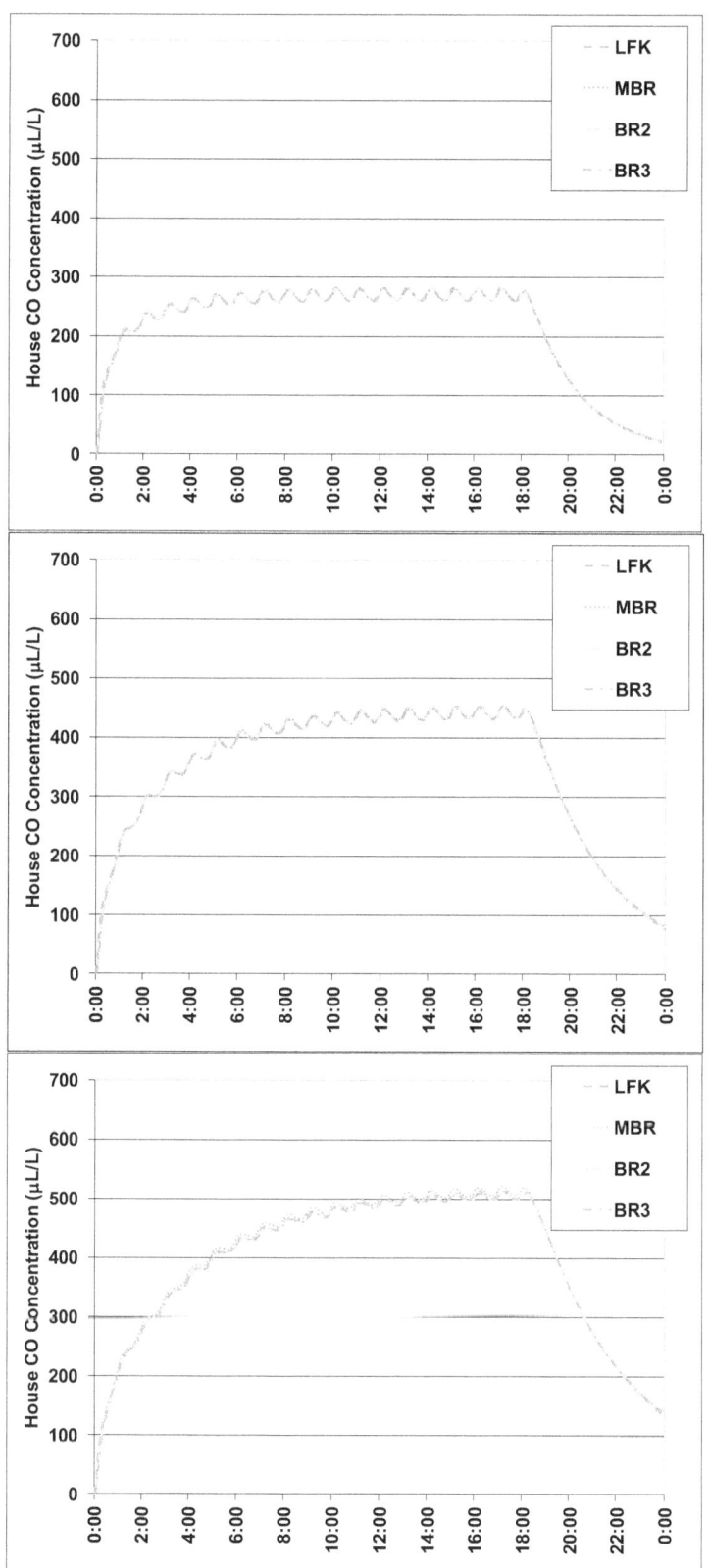

Figure 40 Utility source, closed utility door, cat muffler cases (a) cold (0°C), (b) mild (15°C), (c) hot (30°C) ambient temperature

83

Table 23 Simulation results for utility source cases

Sim ID	Source location	Generator	Utility door	Weather	Utility Room Peak CO Concentration (excluding initial spike) (µL/L)	House Peak CO Concentration, (excluding utility room) (µL/L)
25	utility	SO1 cat	closed	cold	2200	280
26	utility	SO1 cat	closed	mild	2500	460
27	utility	SO1 cat	closed	hot	3500	520
28	utility	SO1 cat	open	cold	1200	330
29	utility	SO1 cat	open	mild	1400	490
30	utility	SO1 cat	open	hot	1500	600
31	utility	SO1 noncat	closed	cold	6000	950
32	utility	SO1 noncat	closed	mild	6600	1500
33	utility	SO1 noncat	closed	hot	7500	1900
34	utility	SO1 noncat	open	cold	3500	1100
35	utility	SO1 noncat	open	mild	3700	1600
36	utility	SO1 noncat	open	hot	4100	2000

Closed bedroom door cases

The cases presented in Tables 22 and 23 all included open bedroom doors, which results in significant mixing between house zones. Selected cases were also simulated with closed bedroom doors to examine the impact of this factor, in part to understand the CO exposure for cases in which occupants sleep with their bedroom doors closed.

Figure 41 presents the simulation results for mild ambient temperature case with Gen SO1 with catmuffler operating in the garage with closed bay door, open house door, and closed bedroom doors. As expected, this case results in far less uniform house zone concentrations than the open bedroom door cases, which range from a peak of 20 µL/L in the BR2 zone to 130 µL/L in the LFK zone.

Table 24 presents the peak house CO concentration for all 6 closed bedroom door cases which show an increase in peak concentration of up to about 30 % in the LFK zone compared to the equivalent cases with open bedroom doors; however, the bedrooms had generally lower concentrations, with the impact ranging from about an 80 % decrease to about a 2 % increase.

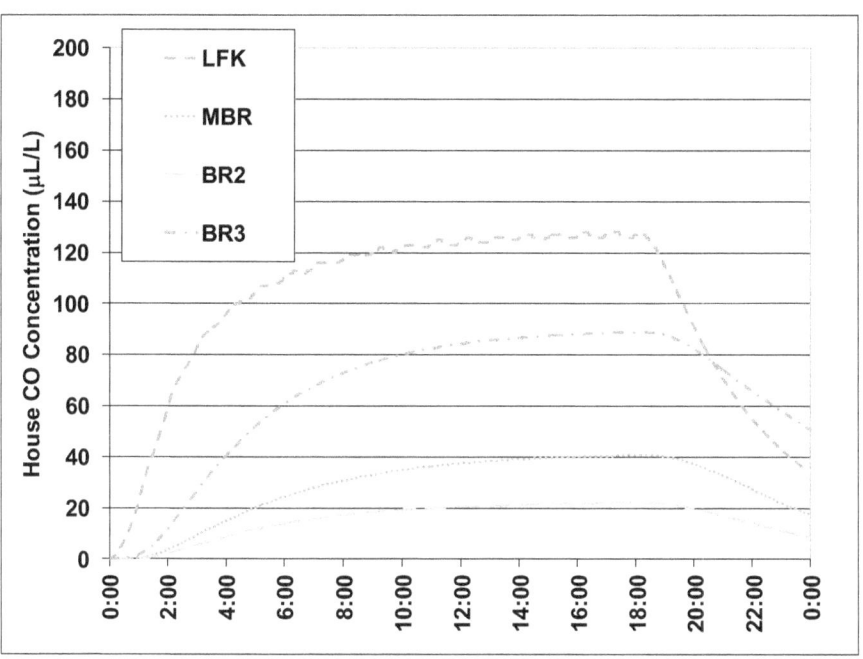

Figure 41 CO concentrations for garage source, closed bay door, open house door, mild ambient temperature, cat muffler with closed bedroom doors

Table 24 Simulation results for closed bedroom door cases

Sim ID	Source location	Generator	Bay or Utility door	House door	Weather	House Peak CO Concentration (excluding utility room) (µL/L)
37	garage	SO1 cat	closed	open	mild	130
38	garage	SO1 cat	closed	closed	cold	40
39	garage	SO1 noncat	open	open	hot	160
40	garage	SO1 noncat	closed	open	mild	450
41	utility	SO1 cat	open	NA	mild	490
42	utility	SO1 noncat	closed	NA	cold	1000

Discussion of Simulation Results

A set of 42 simulations were performed to examine the potential performance of the prototype generator with and without the catalyst in the muffler under a wider range of conditions than studied during the experiments at the test house. All of the simulations were based on the NIST manufactured test house model and the reduced CO emission prototype generator, which was extensively validated as described above. Model parameters that were varied included ambient conditions, CO emission rates, source locations and door positions.

As expected the highest house CO concentrations were found for the generator with noncat muffler operated in the utility room of the house, with indoor CO concentrations reaching 2000 μL/L for one case. Operation of the generator with cat muffler in the utility room substantially reduced CO concentrations, however, they still reached high levels of 280 μL/L to 600 μL/L. The lowest indoor CO concentrations resulted from operation of the cat muffler generator in the garage with peak house CO concentrations as low as 10 μL/L (after an initial spike on start-up). Simulations also showed that, as expected, closed bedroom doors resulted in less uniform indoor concentrations and higher peak concentrations in the LFK zone (up to a 30 % increase) and typically lower peak concentrations in the bedrooms (up to an 80 % decrease).

Conclusion

Under an interagency agreement with CPSC, NIST conducted a series of tests to characterize the indoor time course profiles of CO concentrations resulting from portable generators. These tests include both unmodified as well as modified low CO emission prototype configurations, operating in the attached garage of a home under various use and environmental conditions, and were conducted so CPSC staff could analyze the safety implications of operating the generators under these conditions. NIST used those test data to validate the ability of the CONTAM model to predict CO levels in the garage and the house and to develop an estimate of the uncertainty of its predictions relative to the measured values. NIST also conducted tests with the generators operating in a one-zone shed to derive their CO emission and O_2 consumption rates, to be used as input to their model validation effort as well as for simulations conducted to examine the potential performance of the low CO-emission prototype under a wider range of operating conditions.

Shed Tests

Very limited study has been conducted directly on CO emission and O_2 consumption rates associated with gasoline-powered generators running indoors. Brown (2006) studied the CO emission rates from four different commercially-available generators in an enclosed experimental chamber, where air temperature and air change rate were controlled to provide different operating conditions. However, the air change rates were generally quite high compared with typical residences. Operating a generator in an enclosed space such as a garage or a storage shed, as opposed to a laboratory chamber, will be subject to uncontrolled temperatures and to lower ventilation rates determined by ambient weather conditions. To determine generator CO emission rates, tests were conducted in a single-zone shed on generators operating in the unmodified carbureted configuration as well as in the low CO emission prototype configuration. A literature search did not reveal previous studies on CO emission from generators in real conditions, where O_2 levels can become significantly lower than ambient, and thereby impact CO emission.

For two different unmodified generators (i.e., without CO emission controls), it was found that CO emissions ranged from a low of around 500 g/h at near ambient O_2 levels to a high of nearly 4000 g/h as O_2 approach 17 %. The rates of CO generation and O_2 consumption in these unmodified generators were affected by multiple parameters, with the O_2 level in the space and the actual electrical output of the generator being two of the most important. Tests performed below 17 % O_2 showed a drop off in CO emissions due to poor engine performance under these conditions. Tests of two modified low CO emission prototype generators (i.e., with CO emission controls) showed reductions of CO emissions of over 90 % depending on the specific emission controls and operating conditions and no trend toward higher emission rates was seen as O_2 levels dropped to 18 %.

Garage Tests

A series of tests were also conducted to measure the emission and transport of CO due to operating portable gasoline-powered generators in an attached garage. This series of tests

87

included both unmodified and UA-modified prototype generators operated in the garage attached to NIST's manufactured test house. Testing was conducted under seven different test house/garage configurations to evaluate their impacts on the buildup of CO in the garage and its transport into the different rooms in the house. These configurations included two different garage bay door positions (fully closed or open 0.6 m), two connecting door settings between the garage and the family room (fully closed or open 5 cm), and two house central heating, ventilating, and air conditioning (HVAC) fan settings (on or off). CO concentrations varied widely with peak house CO concentrations ranging from under 10 µL/L to over 10,000 µL/L. As expected, the highest concentrations resulted from operation of the unmodified generator in the garage with the bay door closed and the house access door open. The lowest concentrations resulted from operation of a reduced-emission prototype in the garage with the bay door open and the house access door closed.

These garage tests documented reductions of 85 % to 98 % in CO concentrations due to emissions from two modified, prototype low CO-emission portable generators compared to a "stock" unmodified generator. Note that these results apply to the specific units tested and that other units, modifications and test conditions may produce different results.

Simulations

A set of 42 simulations were performed with the NIST CONTAM model to examine the potential performance of the prototype generators under a wider range of conditions than studied during the experiments at the test house. All of the simulations were based on the NIST manufactured test house model and the tested reduced CO emission prototype generator. Model parameters that were varied included ambient conditions, CO emission rates, source locations and door positions.

An extensive model validation effort was first carried out using results from seven garage tests to establish the ability of the CONTAM model to predict CO levels in the garage and the house and to develop an estimate of its uncertainty of these predictions relative to the measured values. An absolute validation of a complex model, such as CONTAM, is impossible as there are infinite possible building models that can be created by a user. For the cases modeled in this effort, no significant errors in the CONTAM model were identified. The agreement between the measurements and predictions of the O_2 concentrations in the garage and of the average CO calculated for the house zones for the collective set of data was excellent. All of the calculated statistical values met the ASTM D5157 criteria. The agreement, however, was somewhat worse for the garage CO concentrations, with some parameters falling slightly outside the ASTM criteria limits. Overall, the average individual house zone and garage CO concentration predictions and measurements were within about 20 % and 30 % respectively when averaged over all cases.

As expected, the highest house CO concentrations were found for the generator with noncat muffler operated in the utility room of the house, with indoor CO concentrations exceeding 2000 µL/L for some cases. Operation of the generator with cat muffler substantially reduced CO concentrations; however, they still reached levels of 280 µL/L to 600 µL/L. The lowest indoor CO concentrations resulted from operation of the generator with cat muffler in the garage with CO concentrations in the house reaching 10 µL/L to 160 µL/L. Simulations also showed that, as expected, closed bedroom doors resulted in less uniform indoor concentrations and higher peak concentrations in the LFK zone (up to a 30 % increase) and typically lower peak concentrations in the bedrooms (up to an 80 % decrease).

Acknowledgements

This research was supported by the U.S. Consumer Product Safety Commission (CPSC) under interagency agreement CPSC-I-06-0012. The authors would like to acknowledge the contributions of Donald W. Switzer, Janet Buyer, Christopher J. Brown, Sandy Inkster, Han Lim and Susan Bathalon from CPSC, and Gregory T. Linteris and Steven J. Nabinger from NIST.

References

Air Products. 2006. Material Safety Data Sheet (MSDS) for Sulfur Hexafluoride.

ASHRAE. 2009. *Handbook of Fundamentals Chapter 16*. ASHRAE.

ASTM. 2008. *Standard Guide for Statistical Evaluation of Indoor Air Quality Models.* D5157-08 American Society for Testing and Materials.

ASTM. 2006. *Standard test method for determining air change in a single zone by means of a tracer gas dilution.* E741-06 American Society for Testing and Materials.

ASTM. 2010. *Standard Test Method for Determining Air Leakage Rate by Fan Pressurization.* E779-10 American Society for Testing and Materials.

Bassett, M. Infiltration and Leakage Paths in Single Family Houses - a Multizone Infiltration Case Study. (1990) AIVC Technical Note 27, Air Infiltration and Ventilation Centre.

Brown, C. J. 2006. Engine-drive Tools, Phase 1 Test Report for Portable Electric Generators; U.S. Consumer Product Safety Commission: Bethesda, MD.

CFR. 2007. 16 CFR part 1407, *Portable Generators; Final Rule; Labeling Requirements*, Federal Register, 72 FR 1443, January 12, 2007 and January 18, 2007.

CPSC. Portable Generators: Legal Memorandum and staff briefing package for advance notice of proposed rulemaking (ANPR); U.S. Consumer Product Safety Commission: Bethesda, MD, 2006; p 295.

CPSC. Performance and accountability report; U.S. Consumer Product Safety Commission: Bethesda, MD, November, 2007; p 140.

CPSC. Technology Demonstration of a Prototype Low Carbon Monoxide Emission Portable Generator; U.S. Consumer Product Safety Commission: Bethesda, MD, August, 2012; p 413.

Dols WS and Walton GN. *CONTAM User Guide and Program Documentation* (2008) NISTIR 7251, National Institute of Standards and Technology.

Emmerich SJ and Nabinger SJ. *Measurement and Simulation of the IAQ Impact of Particle Air Cleaners in a Single-Zone Building.* (2001) International Journal of HVAC&R Research Vol. 1, No. 7, ASHRAE.

Emmerich SJ. "Validation of Multizone IAQ Modeling of Residential-scale Buildings: A Review" (2001) ASHRAE Transactions Vol. 107.2. 49.

Emmerich, S.J., Gorfain, J.E., Howard-Reed, C., Air and Pollutant Transport from Attached Garages to Residential Living Spaces-Literature Review and Field Test (2003) International Journal of Ventilation, Vol., 2 (3).

Emmerich, S. J., & Persily, A. K. (1996). Multizone modeling of three residential indoor air quality control options. Building Simulation '95.

Emmerich, S. J., Persily, A., & Nabinger, S. 2002. Modeling moisture in residential buildings with a multizone IAQ program. Proceedings of Indoor Air 2002.

Emmerich, S. J., Reed, C. H., & Gupta, A. (2005). *Modeling the IAQ impact of HHI interventions in inner-city housing.* NISTIR 5212, National Institute of Standards and Technology.

Emmerich, S.J., Howard-Reed, C., Nabinger, S.J., Validation of Multizone IAQ Model Predictions for Tracer Gas in a Townhouse (2004) Building Services Engineering Research and Technology, Vol. 25 (4).

Emmerich, S. and Wang, L. 2011. Measured CO Concentrations at NIST IAQ Test House from Operation of Portable Electric Generators in Attached Garage – Interim Report to U.S. Consumer Products Commission.

Hnatov, M. V. 2012. Incidents, Deaths, and In-Depth Investigations Associated with Non-Fire Carbon Monoxide from Engine-Driven Generators and Other Engine-Driven Tools, 1999-2011; U.S. Consumer Product Safety Commission: Bethesda, MD.

Hnatov, M.V. 2011. Non-Fire Carbon Monoxide Deaths Associated with the Use of Consumer Products, 2008 Annual Estimates, U.S. Consumer Product Safety Commission, Bethesda, MD.

Luce, C., Heiser, B., Siemer, M., and B. Smith. 2006. Redesign of the mud buggy to reduce emissions by conversion to propane fuel. In Multi-disciplinary engineering design conference, Rochester, NY.

Nabinger, SJ, AK Persily, and WS Dols. 2010. Impacts of Airtightening Retrofits on Ventilation Rates and Energy Consumption in a Manufactured Home. NIST Technical Note 1673.

Nabinger, SJ and AK Persily. 2008. *Airtightness, Ventilation and Energy Consumption in a Manufactured House: Pre-Retrofit Results.* NISTIR 7478.

Proctor, C. L.; Berger, M. C.; Fournier, D. L.; and S. Roychoudhury. 1987. Sulfur hexafluoride as a tracer for the verification of waste-destruction levels in an incineration process. Florida University.

Stewart, R. D. 1975. The effect of carbon monoxide on humans. Annual review of pharmacology . 15, 409-23.

Taylor, B. N., C. E. Kuyatt and T. National Institute of Standards and. 1994. Guidelines for evaluating and expressing the uncertainty of NIST measurement results. Gaithersburg, MD: U.S. Dept. of Commerce, Technology Administration, National Institute of Standards and Technology.

Wang, L., and Emmerich, S.J. 2010. In situ Experimental Study of Carbon Monoxide Generation by Gasoline-Powered Electric Generator in an Enclosed Space. Journal of the Air & Waste Management Association, Vol 60(12).

Wang, L., Emmerich, S.J., Persily A.K. and C-C Lin. 2012. Carbon Monoxide Generation, Dispersion and Exposure from Indoor Operation of Gasoline-powered Electric Generators under Actual Weather Conditions. Building and Environment 56: 283-290.

Appendix A Uncertainty Analysis of Shed Measurements

This Appendix presents the uncertainty analysis method for calculation of the CO emission and O_2 consumption rates determined from testing the generators in the test shed. Sample calculations are presented in detail.

The uncertainty of A_{out} can be expressed as:

$$A_{out} = \frac{(ln C_{SF_6,t1} - ln C_{SF_6,t2})}{t2 - t1} + \frac{Q_{gen,in}}{V_s} K_{SF_6,d} = A1 + A2$$

So the propagation of uncertainty for A_{out}

$$u_c^2(A_{out}) = u^2(A1) + u^2(A2) + 2\,u(A1, A2)$$

$u(A1)$ is calculated from

$$u(A1) = |A1| \frac{tan(arccos(r(ln C, t)))}{(N-2)^{1/2}}$$

$r(lnC, t)$ is the correlation coefficient of lnC and t.

From our pretests of checking SF_6 decomposition, the covariance of $A1$ and $A2$

$$u(A1, A2) = -0.07341\%$$

$$u^2(A2) = (\frac{K_{SF_6,d}}{V_s})^2 u^2(Q_{gen,in}) + (\frac{Q_{gen,in}}{V_s})^2 u^2(K_{SF_6,d}) + \frac{2K_{SF_6,d}Q_{gen,in}}{V_s^2} u(Q_{gen,in}, K_{SF_6,d})$$

$K_{SF_6,d}$ = -4 %, which were obtained from our pretests of the SF_6 decomposition. To get the uncertainty of $Q_{gen,in}$, we can look at the uncertainty propagation of Eq. (6)

$$u(Q_{gen,in}) = \sqrt{(D_{gen}\frac{RPM}{2})^2 u^2(\eta_{gen})}$$

η_{gen} is 85 % ± 5 % (Luce et al. 2006) so based on Eq. (A-7) of NIST uncertainty guidance(Taylor et al. 1994), we have

$$u(\eta_{gen}) = \frac{5\%}{\sqrt{3}} = 2.89\%$$

Then $u(Q_{gen,in}) = 0.99$ m^3/h. If we assume $u(Q_{gen,in}, K_{SF_6,d}) = 0$, then

$$A2 = \frac{Q_{gen,in}}{V_s} K_{SF_6,d} = -0.0301$$

$$u^2(A2) = (\frac{K_{SF_6,d}}{V_s})^2 u^2(Q_{gen,in}) + (\frac{Q_{gen,in}}{V_s})^2 u^2(K_{SF_6,d})$$

$$u(A2) = 0.0131$$

<u>Example of A_{out} Uncertainty Calculation</u>
To calculate the uncertainty of air change rate in Test 1 of Table 2:

$$u(A1) = 0.18$$

$$u(A1, A2) = -0.07341\%$$

$$u(A2) = 0.0131$$

So $u_c(A_{out}) = \sqrt{u^2(A1) + u^2(A2) + 2u(A1, A2)} = 0.2$. For a confidence level of 95%, select $k = 2$, then $k \cdot u_c(A_{out}) = 0.4$, $A_{out} = 6.5 \pm 0.4$ hr^{-1}

Uncertainty of CO Emission Rate
$$S_{CO} = \rho_{CO,in} A_{out} V_s \frac{C_{CO,t2} - C_{CO,t1} e^{A_{out}(t1-t2)}}{1 - e^{A_{out}(t1-t2)}} = f(\rho_{CO,in}, A_{out}, C_{CO,t2}, C_{CO,t1})$$

The uncertainty propagation for S_{CO} was estimated as

$$u_c^2(S_{CO}) = (\frac{\partial f}{\partial \rho_{CO,in}})^2 u^2(\rho_{CO,in}) + (\frac{\partial f}{\partial A_{out}})^2 u^2(A_{out}) + (\frac{\partial f}{\partial C_{CO,t2}})^2 u^2(C_{CO,t2}) + (\frac{\partial f}{\partial C_{CO,t1}})^2 u^2(C_{CO,t1})$$

where

$$\frac{\partial f}{\partial \rho_{CO,in}} = A_{out} V_s \frac{C_{CO,t2} - C_{CO,t1} e^{A_{out}(t1-t2)}}{1 - e^{A_{out}(t1-t2)}} - \frac{S_{CO}}{\rho_{CO,in}}$$

$$\frac{\partial f}{\partial A_{out}} = \frac{S_{CO}}{A_{out}} + \rho_{CO,in} A_{out} V_s [\frac{(-C_{CO,t1})(t1-t2)e^{A_{out}(t1-t2)}}{1 - e^{A_{out}(t1-t2)}}$$
$$+ \frac{(t1-t2)e^{A_{out}(t1-t2)}(C_{CO,t2} - C_{CO,t1}e^{A_{out}(t1-t2)})}{(1 - e^{A_{out}(t1-t2)})^2}]$$

$$\frac{\partial f}{\partial C_{CO,t2}} = \rho_{CO,in} A_{out} V_s \frac{1}{1 - e^{A_{out}(t1-t2)}}$$

$$\frac{\partial f}{\partial C_{CO,t1}} = -\rho_{CO,in} A_{out} V_s \frac{e^{A_{out}(t1-t2)}}{1 - e^{A_{out}(t1-t2)}}$$

To account for the uncertainty of using averaged density during Δt

$$u(\rho_{CO,in}) = [\frac{1}{N(N-1)} \sum_{i=1}^{N} (\rho_{CO,in,i} - \overline{\rho_{CO,in}})^2]^{1/2}$$

$u(C_{CO,t2})$ and $u(C_{CO,t1})$ were obtained from calibration curves of the *CO* analyzer as follows.

Suppose the regression function from a calibration curve is

$$Y = B1 + B2 \cdot X$$

where X is the expected value, for example, a known concentration of a gas cylinder, and Y is the measured value during calibration.
During a formal test, if Y is the measured raw data

$$X_c = (Y - B1)/B2$$

Where X_c is the corrected data by using the calibration curve, whose uncertainty can be determined from:

$$u(X_c) = \sqrt{C_Y^2 u^2(Y) + C_{B1}^2 u^2(B1) + C_{B2}^2 u^2(B2) + 2C_{B1}C_{B2} u(B1, B2)}$$

where

$$C_Y = \frac{\partial X_c}{\partial Y} = \frac{1}{B2}$$

$$C_{B1} = \frac{\partial X_c}{\partial B1} = -\frac{1}{B2}$$

$$C_{B2} = \frac{\partial X_c}{\partial B2} = -\frac{(Y - B1)}{B2^2}$$

$u(Y) = \dfrac{a}{\sqrt{3}}$, which can be either determined as a Type B uncertainty from a, the accuracy of the analyzer, or as a standard deviation for a Type A uncertainty if multiple observations of Y are available.

$$u(B1, B2) = -\overline{X}u^2(B2)$$

where $\overline{X}$ is the averaged X of the expected value during a calibration.

An Example of CO Emission Rate Uncertainty Calculation

The maximum CO emission rate of Test 1 in Table 2 is 900 g/h. $\dfrac{\partial f}{\partial \rho_{CO,in}} = 0.81$; $\dfrac{\partial f}{\partial A_{out}} = 0.10$; $\dfrac{\partial f}{\partial C_{CO,t2}} = 531.14$; $\dfrac{\partial f}{\partial C_{CO,t1}} = -337.59$; $u(\rho_{CO,in}) = 0.001$

From a calibration curve of the CO analyzer for this test, $u(C_{CO,t1}) = 91.03$ mg/m^3; $u(C_{CO,t2}) = 88.09$ mg/m^3; So the uncertainty of S_{CO} when $S_{CO} = 878$ g/h is

$$u_c(S_{CO}) = \sqrt{(\dfrac{\partial f}{\partial \rho_{CO,in}})^2 u^2(\rho_{CO,in}) + (\dfrac{\partial f}{\partial A_{out}})^2 u^2(A_{out}) + (\dfrac{\partial f}{\partial C_{CO,t2}})^2 u^2(C_{CO,t2}) + (\dfrac{\partial f}{\partial C_{CO,t1}})^2 u^2(C_{CO,t1})}$$
≈ 60 g/hr

For a confidence level of 95 % and $k = 2$, $S_{CO} = 900 \pm 120$ g/h

Uncertainty of O$_2$ Consumption Rate

$$S_{O_2} = \rho_{O_2,in} A_{out} V_s \dfrac{C_{O_2,t2} - C_{O_2,t1} e^{A_{out}(t1-t2)}}{1 - e^{A_{out}(t1-t2)}} - \dfrac{\rho_{O_2,out} A_{out} V_s}{K_{m,\rho}} C_{O_2,out} = f(\rho_{O_2,in}, A_{out}, C_{O_2,t2}, C_{O_2,t1}, K_{m,\rho})$$

The uncertainty propagation for S_{O_2} was

$$u_c^2(S_{O_2}) = (\dfrac{\partial f}{\partial \rho_{O_2,in}})^2 u^2(\rho_{O_2,in}) + (\dfrac{\partial f}{\partial A_{out}})^2 u^2(A_{out}) + (\dfrac{\partial f}{\partial C_{O_2,t2}})^2 u^2(C_{O,t2}) + (\dfrac{\partial f}{\partial C_{O_2,t1}})^2 u^2(C_{O,t1})$$
$$+ (\dfrac{\partial f}{\partial K_{m,\rho}})^2 u^2(K_{m,\rho}) + (\dfrac{\partial f}{\partial S_{gen}})^2 u^2(S_{gen})$$

where

$$\dfrac{\partial f}{\partial \rho_{O_2,in}} = A_{out} V_s \dfrac{C_{O_2,t2} - C_{O_2,t1} e^{A_{out}(t1-t2)}}{1 - e^{A_{out}(t1-t2)}}$$

$$\frac{\partial f}{\partial A_{out}} = \rho_{O_2,in} V_s \frac{C_{O_2,t2} - C_{O_2,t1} e^{A_{out}(t1-t2)}}{1 - e^{A_{out}(t1-t2)}} + \rho_{O_2,in} A_{out} V_s \left[\frac{(-C_{O_2,t1})(t1-t2) e^{A_{out}(t1-t2)}}{1 - e^{A_{out}(t1-t2)}} \right.$$

$$+ \left. \frac{(t1-t2) e^{A_{out}(t1-t2)}(C_{O_2,t2} - C_{O_2,t1} e^{A_{out}(t1-t2)})}{(1 - e^{A_{out}(t1-t2)})^2} \right] - \frac{\rho_{O_2,out} V_s C_{o_2,out}}{K_{m,\rho}}$$

$$\frac{\partial f}{\partial C_{O_2,t2}} = \rho_{O_2,in} A_{out} V_s \frac{1}{1 - e^{A_{out}(t1-t2)}}$$

$$\frac{\partial f}{\partial C_{O_2,t1}} = -\rho_{O_2,in} A_{out} V_s \frac{e^{A_{out}(t1-t2)}}{1 - e^{A_{out}(t1-t2)}}$$

$$\frac{\partial f}{\partial K_{m,\rho}} = \frac{\rho_{O_2,out} A_{out} V_s}{K^2_{m,\rho}} C_{o_2,out}$$

$$u(\rho_{O_2,in}) = \left[\frac{1}{N(N-1)} \sum_{i=1}^{N} (\rho_{O_2,in,i} - \overline{\rho_{O_2,in}})^2 \right]^{1/2}$$

$$u(A_{out}) = [u^2(A1) + u^2(A2) + 2 u(A1, A2)]^{1/2}$$

$u(C_{O_2,t2})$ and $u(C_{O_2,t1})$ can be obtained using the same method as $u(C_{CO,t2})$ and $u(C_{CO,t1})$.

$$K_{m,\rho} = \frac{\rho_{air,out}}{\rho_{m,in}}$$

$$u(K_{m,\rho}) = \left[\frac{1}{N(N-1)} \sum_{i=1}^{N} (K_{m,\rho,i} - \overline{K_{m,\rho}})^2 \right]^{1/2}$$

Example of O_2 Consumption Rate Uncertainty Calculation

The maximum O_2 consumption rate of Test 1 in Table 2 is 4800 g/hr. $\frac{\partial f}{\partial \rho_{O_2,in}} = 55.79$;

$\frac{\partial f}{\partial A_{out}} = -0.69$; $\frac{\partial f}{\partial C_{O_2,t2}} = 743.30$; $\frac{\partial f}{\partial C_{O_2,t1}} = -395.40$; $\frac{\partial f}{\partial K_{m,\rho}} = 65.41$; $u(\rho_{O_2,i}) = 4.6 \times 10^{-4}$; $u(C_{O_2,t2}) = u(C_{O_2,t1}) = 0.01\%$

When neglecting the added fuel from the generator to the shed, $S_{gen} = 0$

96

$$u_c^2(S_{O_2}) = (\frac{\partial f}{\partial \rho_{O_2,in}})^2 u^2(\rho_{O_2,in}) + (\frac{\partial f}{\partial A_{out}})^2 u^2(A_{out}) + (\frac{\partial f}{\partial C_{O_2,t2}})^2 u^2(C_{O_2,t2}) + (\frac{\partial f}{\partial C_{O_2,t1}})^2 u^2(C_{O_2,t1})$$

$$+ (\frac{\partial f}{\partial K_{m,\rho}})^2 u^2(K_{m,\rho})$$

$u_c(S_{O_2}) \approx 170$ g/h. For a confidence level of 95% and $k = 2$, $S_{O_2} = 4800 \pm 340$ g/h

Appendix B Summary of Instrument Calibrations

This table summarizes the calibrations of the CO and O_2 analyzers covering the testing periods included in this report. The table includes the date of the calibrations, the standard error for each instrument channel for each calibration, and the average standard error and the average standard error relative to the full scale for each device based on all of the calibrations. Not all analyzer channels were calibrated on each date due to instrument failure or other issues. Table 3 in the report describes which instrument was used for each test, and the instruments are described in the Instrumentation section of the main body. For comparison, the manufacturer's stated accuracy for all of these analyzers is 1 % of full scale.

Date	N2 O_2 std error (%)	N1 O_2 std error (%)	N2 hi CO std error (%)	N2 lo CO std error (%)	N1 hi CO std error (%)	N1 lo CO std error (%)	Nova3 CO std error (ppm)	TE CO std error (ppm)	RM CO std error (ppm)
3/17/2008	0.0105	0.0191	0.0160	0.0036	0.0096	0.0056	NA	NA	
4/17/2008	0.0203	0.0243	NA	NA	NA	0.0094	26.3	NA	
4/21/2008	0.482	0.0290	0.0107	0.0033	0.0033	0.0072	23.4	NA	NA
4/29/2008	0.0317	0.0299	0.0090	0.0035	0.0026	0.0031	18.1	NA	NA
5/5/2008	0.0210	0.0344	0.0052	0.0035	0.0028	0.0056	18.1	NA	NA
5/13/2008	0.0255	0.0794	0.0397	0.0229	0.0074	0.0094	10.8	23.0	NA
5/21/2008	0.0192	0.0305	0.0026	0.0059	0.0062	0.0094	26.0	18.0	NA
6/2/2008	0.0551	0.0225	0.0108	0.0074	0.0065	0.0035	NA	NA	NA
6/10/2008	0.0140	0.0298	0.0086	0.0108	0.0081	0.0155	44.4	NA	NA
3/17/2010	0.239	NA	0.0090	NA	0.0070	0.0045	14.4	NA	NA
4/9/2010	0.0543	NA	0.0029	0.0065	0.0091	0.0067	13.8	NA	0.387
4/28/2010	0.0625	NA	0.0056	0.0004	0.0028	0.0003	11.0	NA	NA
5/12/2010	0.0798	NA	0.0088	0.0253	0.0028	0.0057	6.87	NA	3.22
5/27/2010	0.0745	NA	0.0144	0.0215	0.0076	0.0225	11.6	NA	4.62
7/1/2010	0.0443	NA	0.0086	0.0123	0.0447	0.0056	15.9	NA	6.36
Average of all calibrations	0.0822	0.0332	0.0108	0.0098	0.0086	0.0076	18.5	20.5	3.65
Percent of full scale	0.33	0.13	0.36	1.08	0.29	0.85	1.03	2.27	0.41

Appendix C Additional Garage Tests

Table C1 provides a complete listing of all tests of the portable generators conducted in the test house garage. The results of these tests were not included in the interim report (Emmerich and Wang 2011), but are provided here to support additional analysis of the results. Plots of the test results that were not included in the main body of this report are included in Figures C1 through C27. The tests are grouped by test house configuration (see Table C2 for description of test house configurations).

Table C1 Additional Tests Conducted in Attached Garage

Generator	Test ID	Date	Load Profile	Test house Config	Outdoor Temp (°C)	Wind speed (m/s)	Approx. Run Time (h)	Garage Peak CO Concentration (μL/L)	Lowest Garage O_2 Concentration (%)	Automoatic shut-off activated	Notes
unmod GenX	B	4/22/08	cyclic profile	1	20.1	6.5	3	19,500	17.1	NA	Figure 13
unmod GenX	E	5/1/08	cyclic profile	1	13.3	1.8	2	13,100	17.5	NA	Figure C1
modGenX with cat	O	4/2/10	cyclic profile	1	22	6.5	4.5	3000	19.4	NA	Figure 14
SO1 with catmuffler and algorithm disabled	N	4/1/10	cyclic profile	1	19.9	6.3	2	300	19.4	NA	Figure 15
SO1 with catmuffler and algorithm enabled	L	3/25/10	cyclic profile	1	21.3	8.4	0.5	420	20.5	Yes	Not plotted due to instrument error.
SO1 with catmuffler and algorithm enabled	M	3/31/10	cyclic profile	1	15.1	10.0	0.8	370	20.3	Yes	Figure C2
Gen B	P	4/6/10	cyclic profile	1	34.0	9.0	2	5500	19.8	NA	Figure C3
Gen B	AU	7/9/10	cyclic profile	1	32.0	6.5	1	3200	20.0	NA	Figure C4
Gen B	AG	5/10/10	500 W	1	17.5	8.6	2	5100	20.0	NA	Figure C5

Generator	Test ID	Date	Load Profile	Test house Config	Outdoor Temp (°C)	Wind speed (m/s)	Approx. Run Time (h)	Garage Peak CO Concentration ($\mu L/L$)	Lowest Garage O_2 Concentration (%)	Automoatic shut-off activated	Notes
SO1 with noncat muffler and algorithm enabled	AP	5/21/10	500 W	1	23.7	6.9	0.2	270	20.9	Y	Figure C6
SO1 with noncat muffler and algorithm enabled	AV	7/9/10	500 W	1	35.1	6.3	2 (manually shut off)	270	20.3	No	Figure C7
Gen B	AI	5/14/10	5500 W	1	20.8	7.5	1.1	5500	19.6	NA	Figure C8
SO1 with noncat muffler and algorithm disabled	AR	5/21/10	5500 W	1	32.8	6.7	4	920	19.0	NA	Figure C9
SO1 with noncat muffler and algorithm disabled	AS	6/10/10	5500 W	1	27.6	8.5	4	890	19.0	NA	Figure C10
SO1 with noncat muffler	AL	5/19/10	5500 W	1	16.9	7.9	0.7 (manually shut off)	810	20.1	No	Figure C11

Generator	Test ID	Date	Load Profile	Test house Config	Outdoor Temp (°C)	Wind speed (m/s)	Approx. Run Time (h)	Garage Peak CO Concentration (µL/L)	Lowest Garage O_2 Concentration (%)	Automoatic shut-off activated	Notes
and algorithm enabled											
SO1 with noncat muffler and algorithm enabled	AM	5/20/10	5500 W	1	24.0	7.7	7 min	720	21.0	Y	Figure C12
SO1 with noncat muffler and algorithm enabled	AN	5/20/10	2500 W	1	28.5	6.6	13 min	400	21.1	Y	Figure C13
unmod GenX	F	5/6/08	cyclic profile	2	22.8	7.7	4	1,500	20.5	NA	Figure 16
unmod GenX	H	5/14/08	cyclic profile	2	23.2	4.0	4	1,200	20.8	NA	Figure C14
modGenX	R	4/12/10	cyclic profile	2	19.9	6.7	4	30	20.7	NA	Figure 17
SO1 with catmuffler and algorithm disabled	T	4/14/10	cyclic profile	2	12.7	6.9	3	300	20.7	NA	Figure 18
SO1 with catmuffler and	S	4/13/10	cyclic profile	2	10.5	6.4	0.9	210	21.0	Y	Figure C15

Generator	Test ID	Date	Load Profile	Test house Config	Outdoor Temp (°C)	Wind speed (m/s)	Approx. Run Time (h)	Garage Peak CO Concentration (µL/L)	Lowest Garage O$_2$ Concentration (%)	Automoatic shut-off activated	Notes
algorithm enabled											
Gen B	Q	4/7/10	cyclic profile	2	30.0	7.7	4	480	21.0	NA	Figure C16
unmod GenX	I	5/15/08	cyclic profile	3	22.8	7.4	4	18,600	17.5	NA	Figure 19
unmod GenX	A	4/18/08	cyclic profile	3	24.5	2.5	1.5	9,200	18.6	NA	Figure C17
unmod GenX	C	4/24/08	cyclic profile	3	22.2	2.9	4	21,000	17.0	NA	Figure C18
SO1 with noncat muffler and algorithm disabled	Z	5/5/10	cyclic profile	3	28.3	6.7	4.75	630	19.5	NA	Figure 20
SO1 with noncat muffler and algorithm enabled	AA	5/6/10	cyclic profile	3	27.6	9.0	1.6	1000	20.0	Y	Figure C19
unmod GenX	J	5/21/08	cyclic profile	4	18.2	9.6	2.25	21,300	16.0	NA	Figure 21
SO1 with catmuffler and algorithm disabled	W	4/29/10	cyclic profile	4	17.8	9.5	6	960	18.2	NA	Figure 22

Generator	Test ID	Date	Load Profile	Test house Config	Outdoor Temp (°C)	Wind speed (m/s)	Approx. Run Time (h)	Garage Peak CO Concentration (μL/L)	Lowest Garage O_2 Concentration (%)	Automoatic shut-off activated	Notes
SO1 with catmuffler and algorithm enabled	X	4/30/10	cyclic profile	4	24.7	7.1	0.6	730	19.8	Y	Figure C20
unmod GenX	D	4/30/08	cyclic profile	5	12.2	8.2	2	23,000	below 16	NA	Figure 23
SO1 with noncat muffler and algorithm disabled	AH	5/13/10	cyclic profile	5	15.6	6.5	5	2,300	17.8	NA	Figure 24
SO1 with noncat muffler and algorithm enabled	AB	5/7/10	cyclic profile	5	21.6	6.6	1	1500	19.2	Y	Figure C21
SO1 with noncat muffler and algorithm enabled	AQ	5/21/10	cyclic profile	5	27.7	6.7	0.8	660	19.4	Y	Figure C22
unmod GenX	G	5/7/08	cyclic profile	6	25.1	7	2	1,100	20.5	NA	Figure 25
SO1 with catmuffler	U	4/22/10	cyclic profile	6	20.4	7.8	2	260	20.9	NA	Figure 26

Generator	Test ID	Date	Load Profile	Test house Config	Outdoor Temp (°C)	Wind speed (m/s)	Approx. Run Time (h)	Garage Peak CO Concentration (μL/L)	Lowest Garage O$_2$ Concentration (%)	Automoatic shut-off activated	Notes
and algorithm disabled											
SO1 with catmuffler and algorithm enabled	Y	4/30/10	cyclic profile	6	28.3	7.8	0.5	250	21	Y	Figure C23
unmod GenX	K	5/23/08	cyclic profile (high to low)	7	13.8	7	>2	680	20.4	NA	Figure 27
SO1 with noncat muffler and algorithm disabled	V	4/23/10	cyclic profile (high to low)	7	15.8	6.5	>2	430	20.9	NA	Figure 28
SO1 with noncat muffler and algorithm enabled	AC	5/7/10	cyclic profile	7	24.6	6.5	10 min	520	21	Y (but not confirmed that it was algorithm)	Not plotted as test was only 10 min
SO1 with noncat muffler and algorithm	AD	5/7/10	cyclic profile	7	24.1	7.1	6 min	340	21	Y	Not plotted as test was only 6 min

Generator	Test ID	Date	Load Profile	Test house Config	Outdoor Temp (°C)	Wind speed (m/s)	Approx. Run Time (h)	Garage Peak CO Concentration (µL/L)	Lowest Garage O_2 Concentration (%)	Automoatic shut-off activated	Notes
enabled											
SO1 with noncat muffler and algorithm enabled	AW	7/9/10	cyclic (high to low)	7	29.3	6.3	1	290	21.1	Y	Figure C27
SO1 with noncat muffler and algorithm enabled	AO	5/20/10	cyclic profile	7	30.6	6.4	1.1	110	20.9	Y	Figure C25
Gen B	AJ	5/19/10	5.5 kW	8	15.8	7.1	3	190	21.1	NA	Figure C26
SO1 with noncat muffler and algorithm enabled	AK	5/19/10	5.5 kW	8	16.9	8.2	1.4	200	21.1	Y	Figure C27

Table C2 House Configurations

Test house Configuration	Garage bay door	Garage to house entry door	HVAC fan
1	Closed	Open 5 cm	OFF
2	Open 0.6 m	Closed	Off
3	Closed	Open 5 cm	On
4	Closed	Closed	On
5	Closed	Closed	Off
6	Open 0.6 m	Open 5 cm	On
7	Open 0.6 m	Open 5 cm	Off
8	Fully Open	Open 5 cm	Off

Figures C1a, C1b, and C1c show the results for Test E, which was a 2 h test of unmod Gen X in Configuration 1 (garage bay door closed, garage access door to house open nominally 5 cm, and the house central HVAC fan off) with the cyclic load profile (see Table 1 in the body of the report). Since it was a two hour test, the hourly cyclic load profile in Table 1 was repeated two times. At the end of the second cycle, the generator was stopped, and the garage was mechanically vented.

Figure C1a CO and O_2 concentrations in the garage and measured load for Test E (unmod Gen X, Configuration 1)

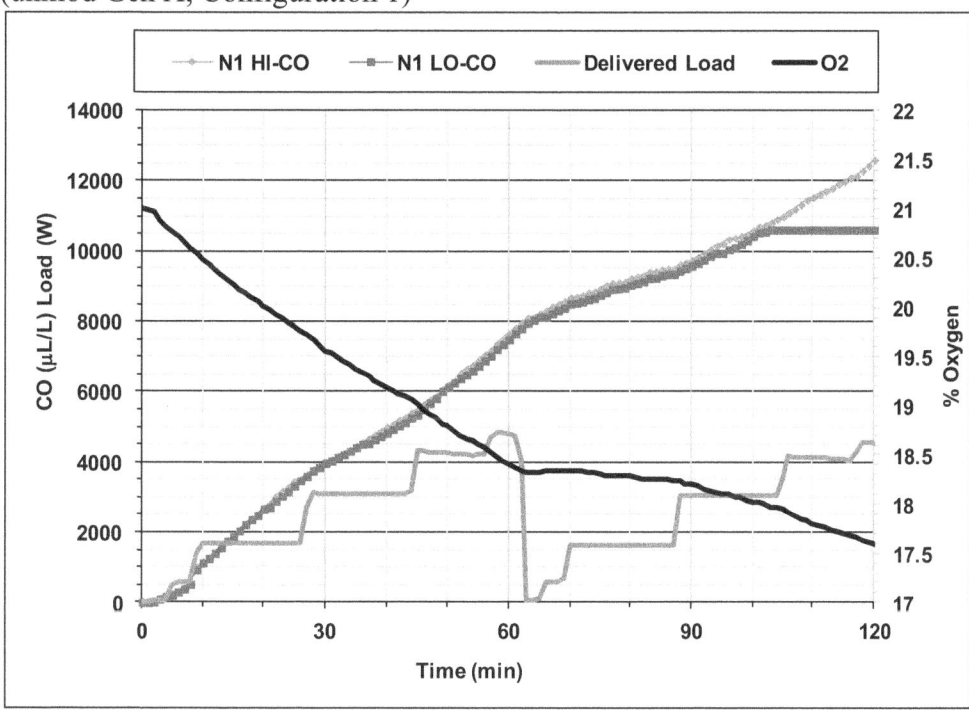

Figure C1a shows the concentration of CO in the garage reached a peak of over 13,000 μL/L (note that μL/L are equivalent to the commonly used unit for CO concentration of ppm_v) and the volume fraction of O_2 dropped by 3.5 % to nearly 17.5 % when the generator was stopped. It also shows that in the first load cycle, the delivered electrical load was less than the load bank settings for the two highest loads in the cycle, 4500 W and 5500 W, which were applied when the oxygen was below 19 %. As the

107

oxygen continued to drop in the subsequent load cycles, the delivered power for these load points decreased further.

Figures C1b and C1c show the CO concentration in six rooms of the test house (see Figure 2 for room locations) as measured on the 'ppm range' (where the CO concentration plot plateaus at the instrument's 2000 μL/L limit) and 'high range' CO instruments, respectively. The CO reached a peak concentration of over 6000 μL/L in the family room, with peak concentrations in the other rooms ranging from about 4000 μL/L to 6000 μL/L. The results of Test E were similar to those of Test B, described in the body of the report, which was the same generator under the same house configuration.

Figure C1b CO (ppm range) concentrations in the house for Test E (unmod Gen X, Configuration 1)

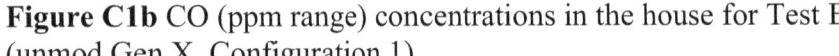
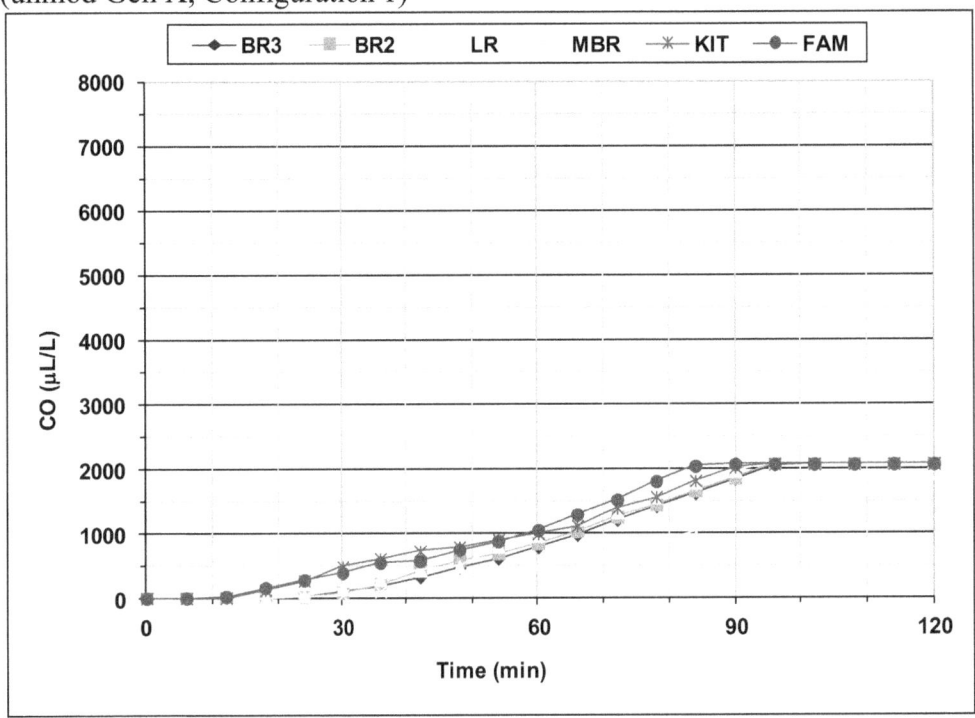

Figure C1c CO (high range) concentrations in the house for Test E (unmod Gen X, Configuration 1)

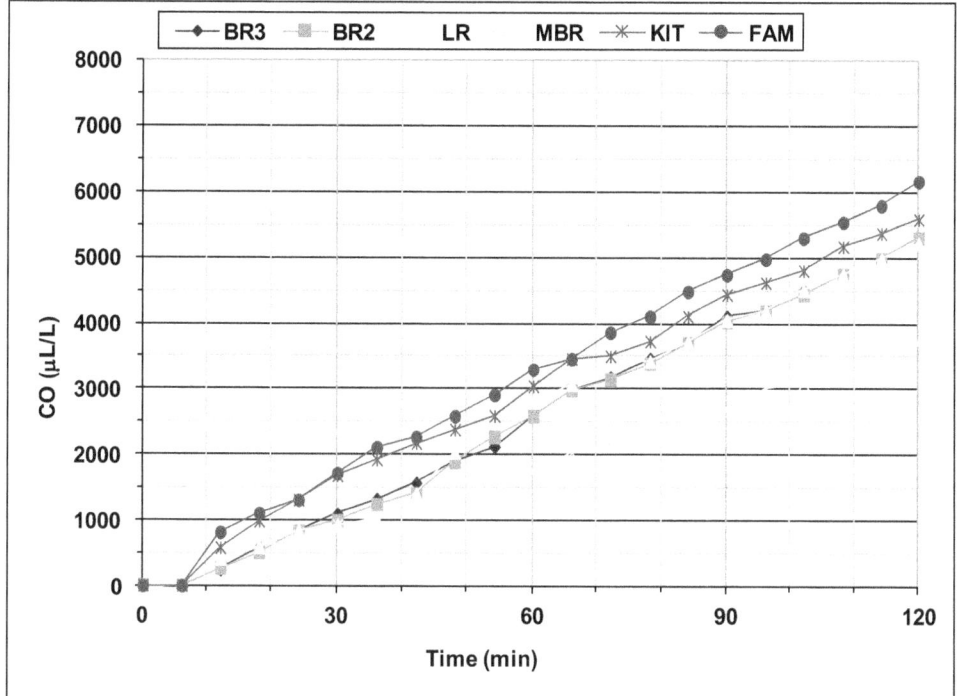

Figures C2a and C2b show the results for Test M, which was a shut-off test of Gen SO1with the shutoff algorithm enabled under test house configuration 1 and the cyclic load. The algorithm shut off the generator after approximately 0.8 h. A natural decay period of about 0.5 h was included after the generator was stopped, followed by mechanical venting. The results shown in Figures C2a and C2b for Test M, prior to generator shut-off, are similar to those in Figure 15 for Test N in the body of the report (a test of the same generator and house configuration but without the shut-off activated).

Figure C2a CO and O_2 concentrations in the garage and measured load for Test M (Gen SO1, Configuration 1)

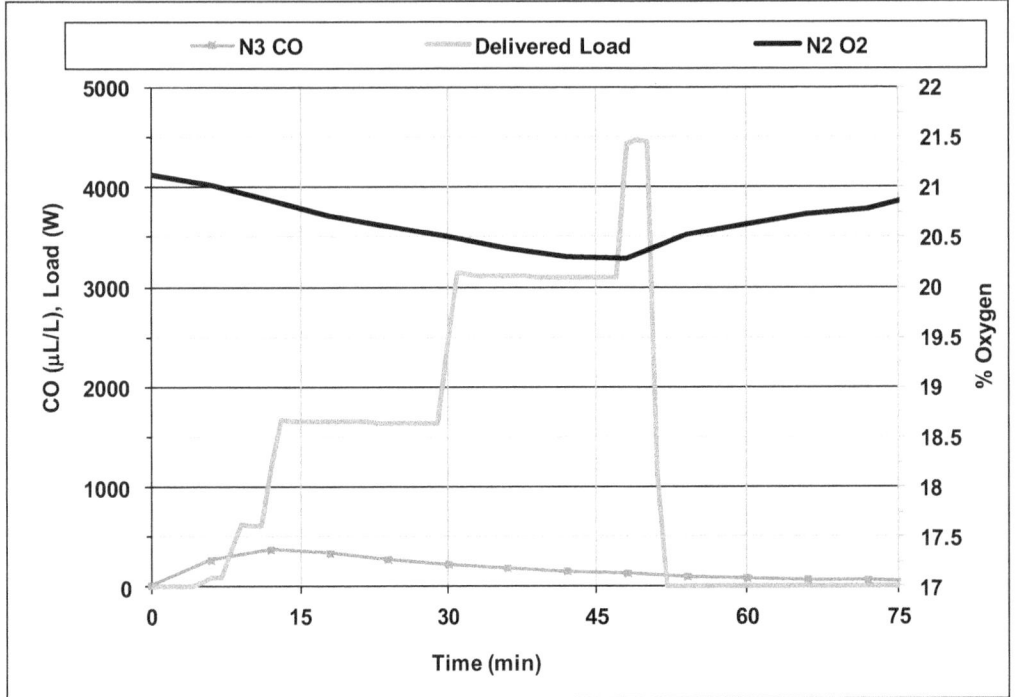

Figure C2b CO concentrations in the house for Test M (Gen SO1, Configuration 1)

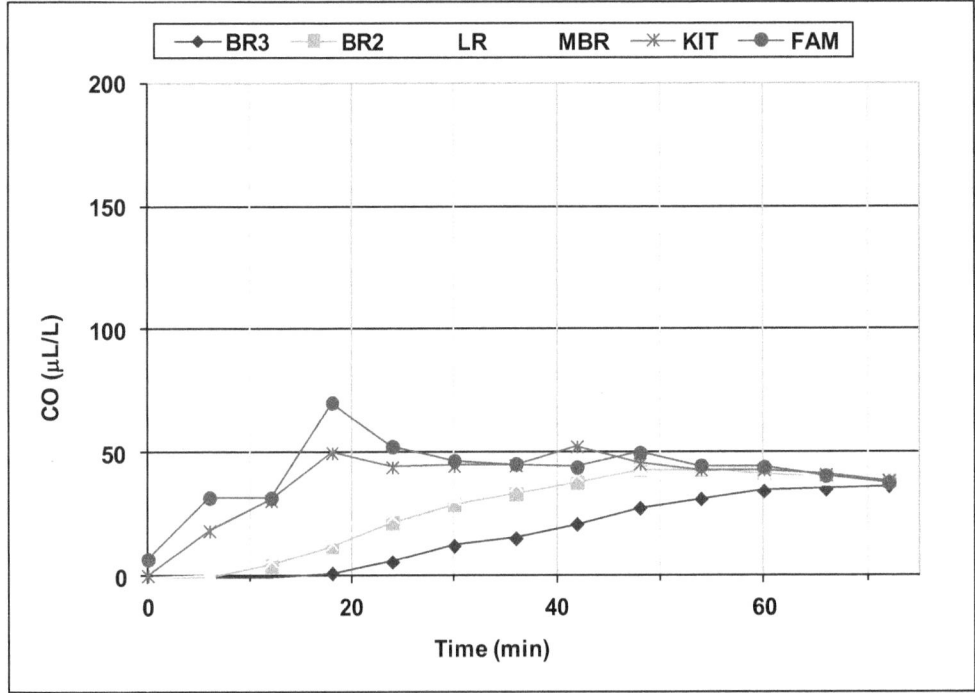

Figures C3a, C3b (the CO concentration exceeded the limit of the ppm range instrument), and C3c show the results for Test P, which was a two hour test of Gen B in Configuration 1 with the cyclic load. Figure C3a shows the concentration of CO in the garage reached a peak of over 5500 μL/L and the volume fraction of O₂ in the garage dropped to about 19.8 % when the generator was stopped.

Figure C3a CO and O₂ concentrations in the garage and measured load for Test P (Gen B, Configuration 1)

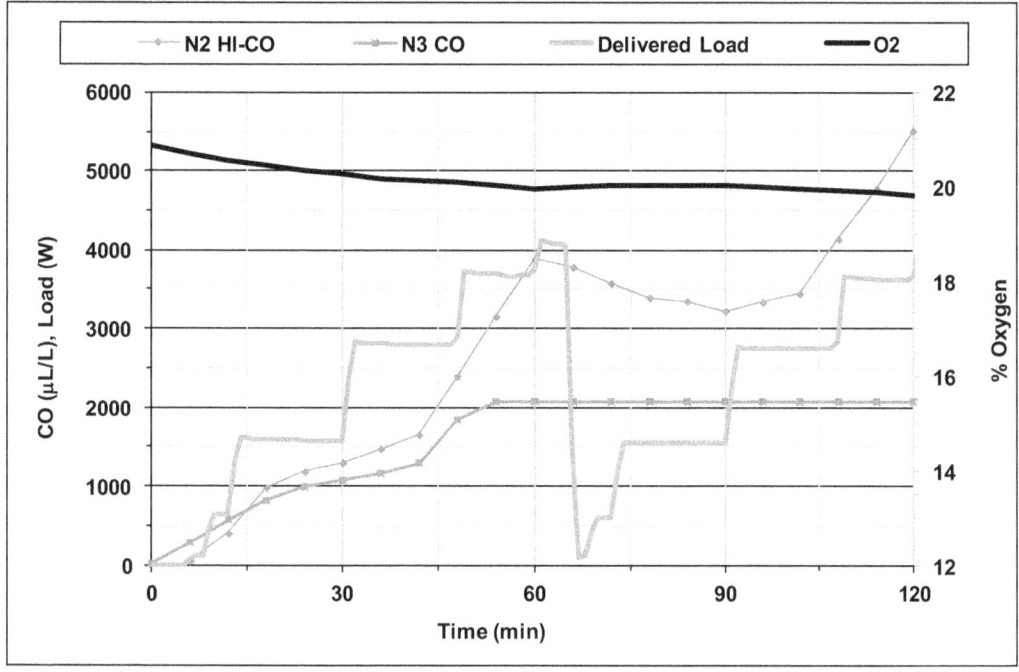

Figure C3b CO (ppm range) concentrations in the house for Test P (Gen B, Configuration 1)

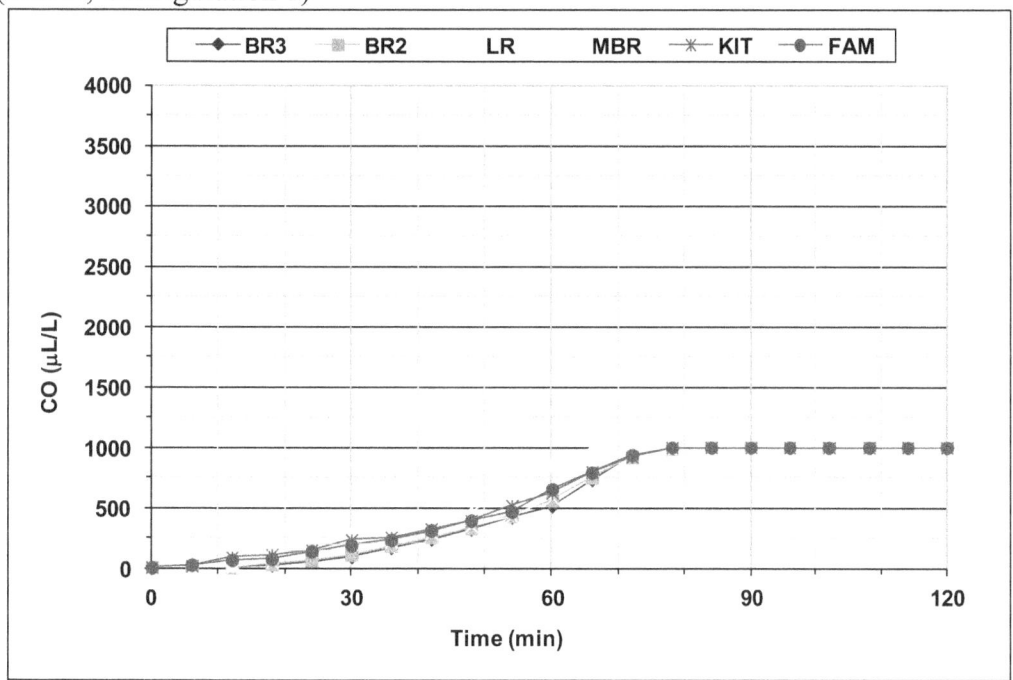

Figure C3c CO (high range) concentrations in the house for Test P (Gen B, Configuration 1)

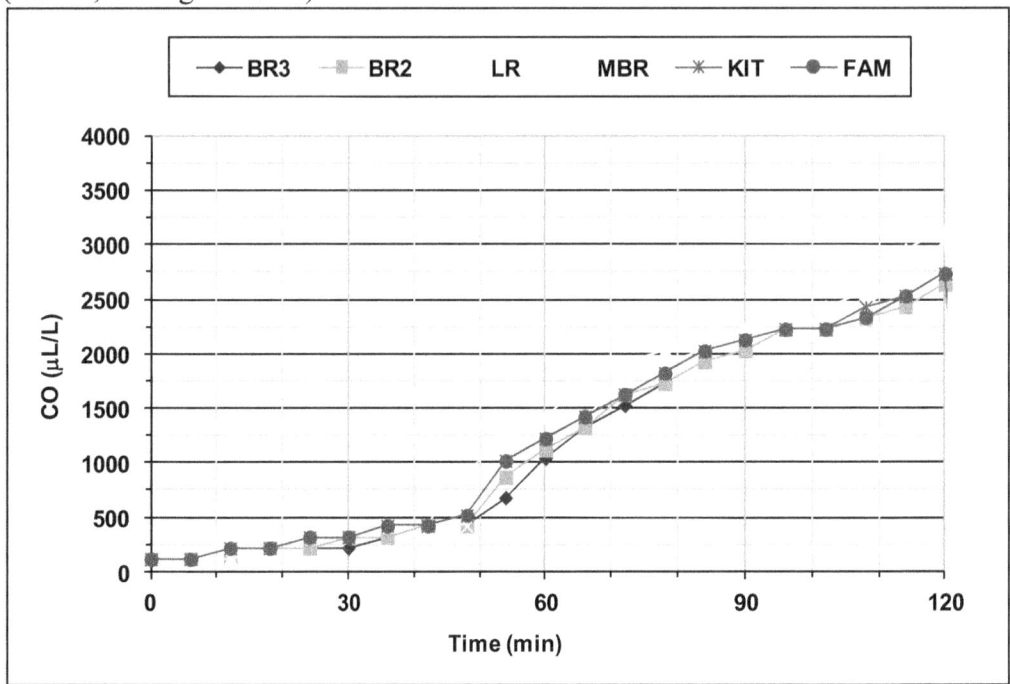

Figures C4a and C4b show the results for Test AU, which was a one hour test of Gen B in Configuration 1 with the cyclic load. Figure C4a shows the concentration of CO in the garage reached a peak of over 3000 μL/L and the volume fraction of O_2 in the garage dropped to about 20 % when the generator was stopped.

Figure C4a CO and O_2 concentrations in the garage and measured load for Test AU (Gen B, Configuration 1)

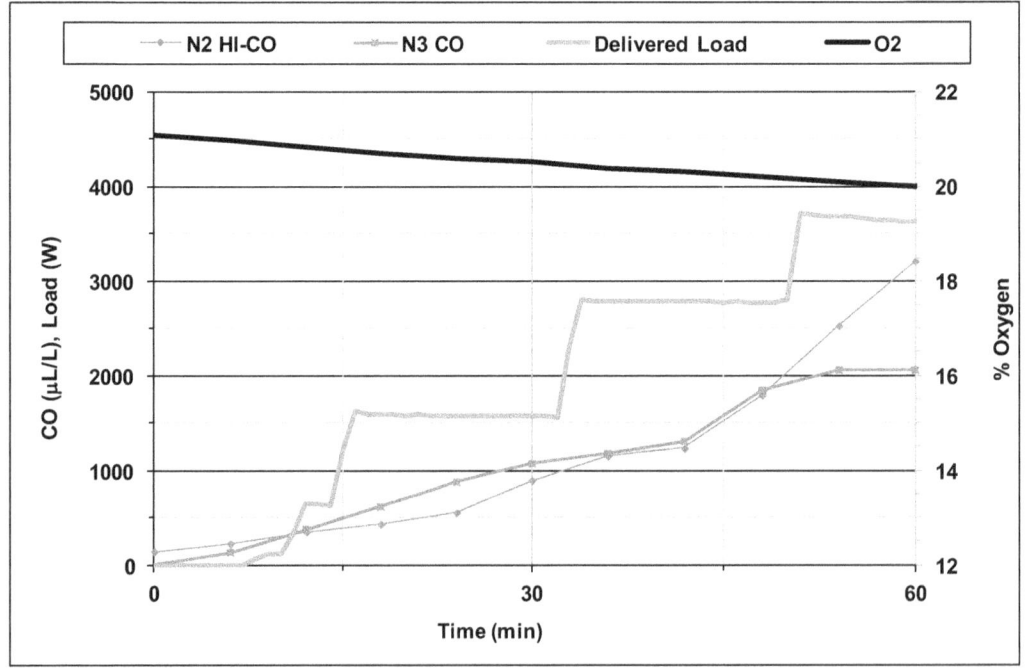

Figure C4b CO (ppm range) concentrations in the house for Test AU (Gen B, Configuration 1)

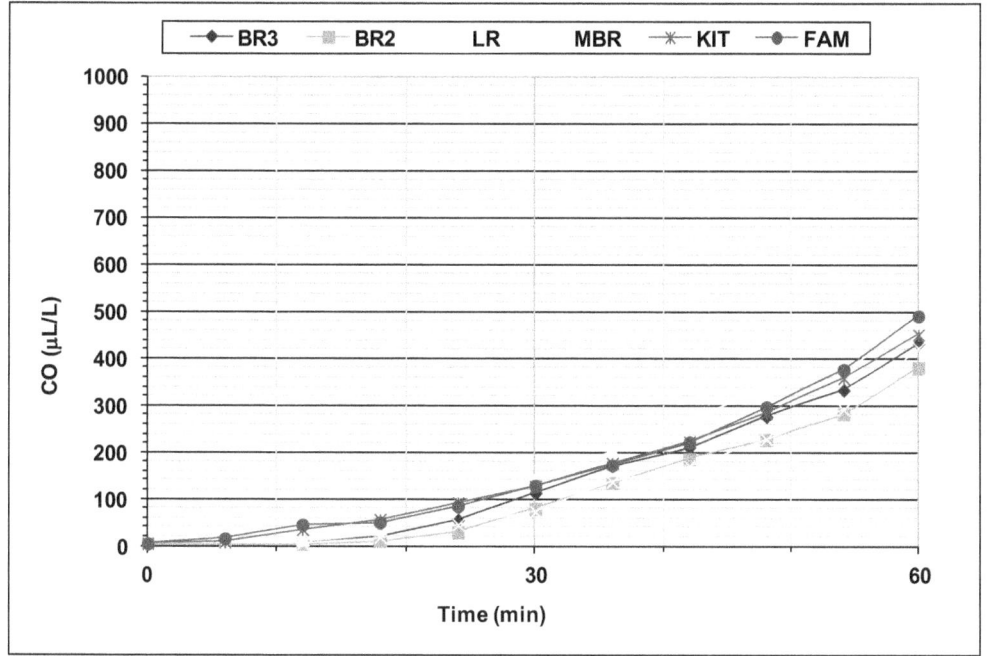

Figures C5a and C5b show the results for Test AG, which was a 2 h test of Gen B in Configuration 1 with a constant 500 W load. The concentration of CO in the garage reached a peak of over 5000 µL/L and the volume fraction of O_2 in the garage dropped to about 20 % when the generator was stopped.

Figure C5a CO and O_2 concentrations in the garage and measured load for Test AG (Gen B, Configuration 1)

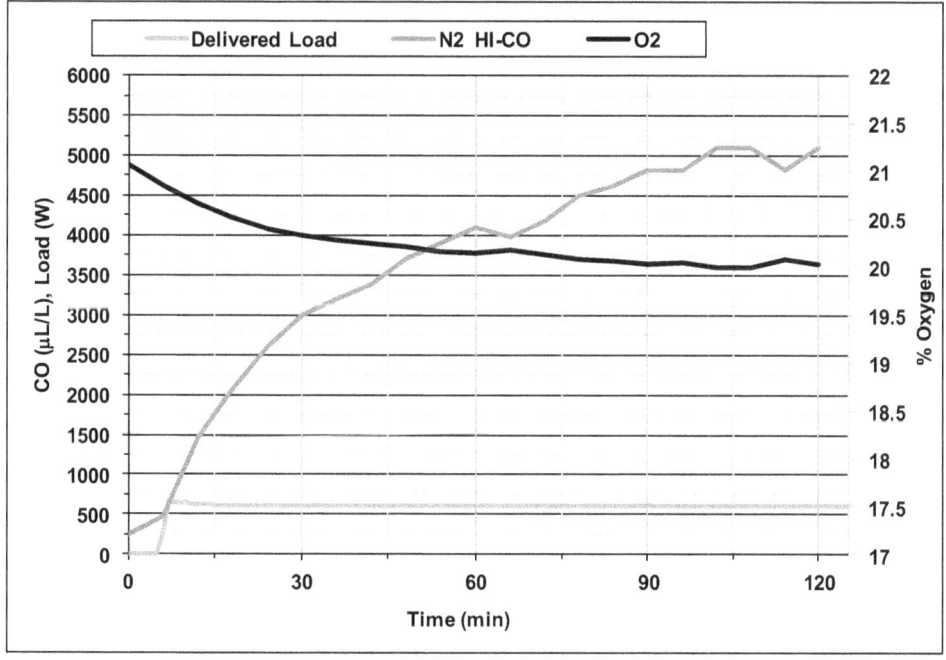

Figure C5b CO concentrations in the house for Test AG
(Gen B, Configuration 1)

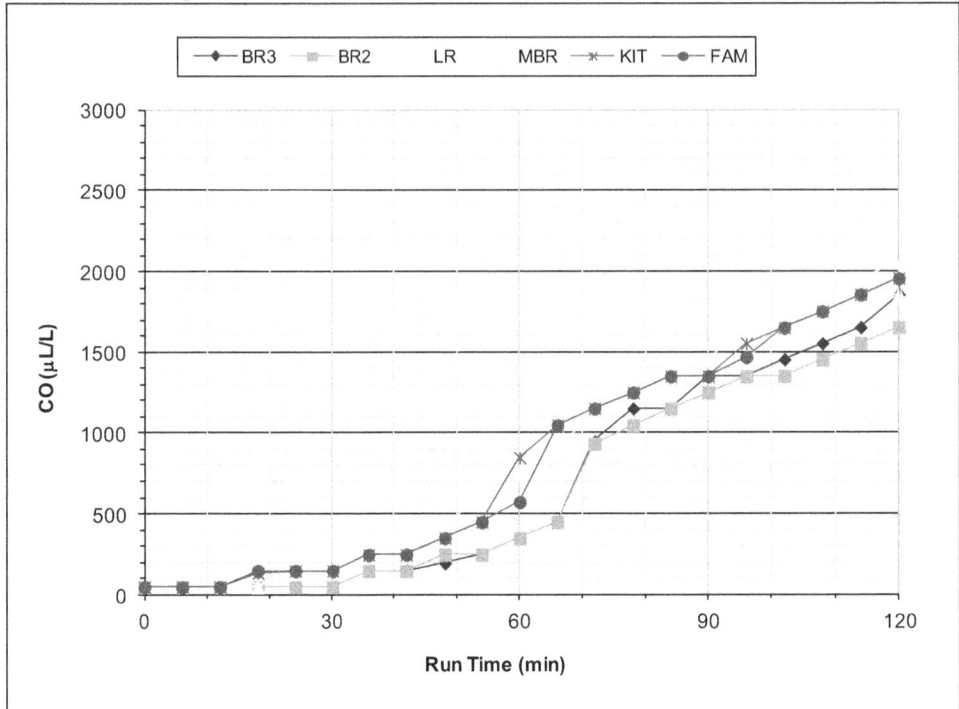

Figures C6a and C6b show the results for Test AP, which was a shut-off test of Gen SO1with the shutoff algorithm enabled under test house configuration 1 and a constant 500 W load. The algorithm shut off the generator after approximately 10 min. A natural decay period of about 20 min was included after the generator was stopped, followed by mechanical venting.

Figure C6a CO and O_2 concentrations in the garage and measured load for Test AP (Gen SO1, Configuration 1)

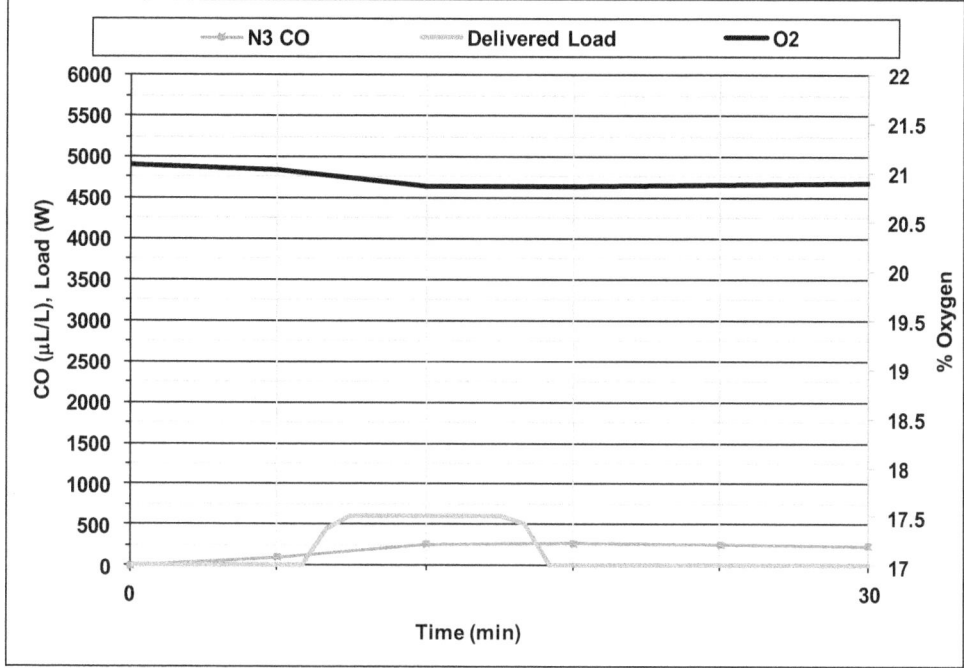

Figure C6b CO concentrations in the house for Test AP (Gen SO1, Configuration 1)

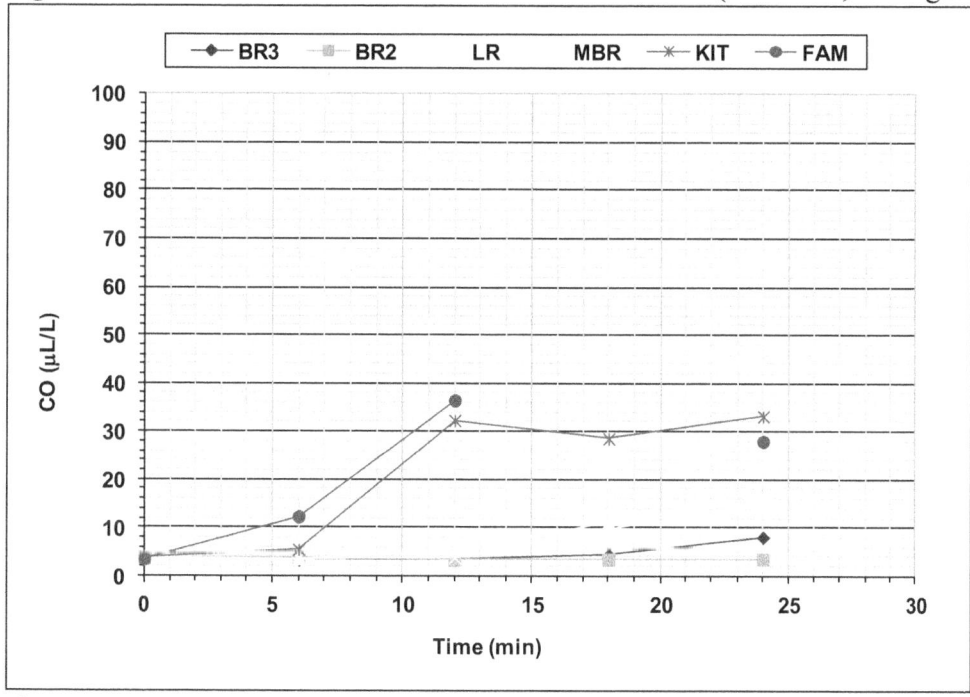

Figures C7a and C7b show the results for Test AV, which was also a shut-off test of Gen SO1 with the shutoff algorithm enabled under test house configuration 1 and a constant 500 W load. The algorithm did not shut off the generator so it was stopped manually after 2 h.

Figure C7a CO and O_2 concentrations in the garage and measured load for Test AV (Gen SO1, Configuration 1)

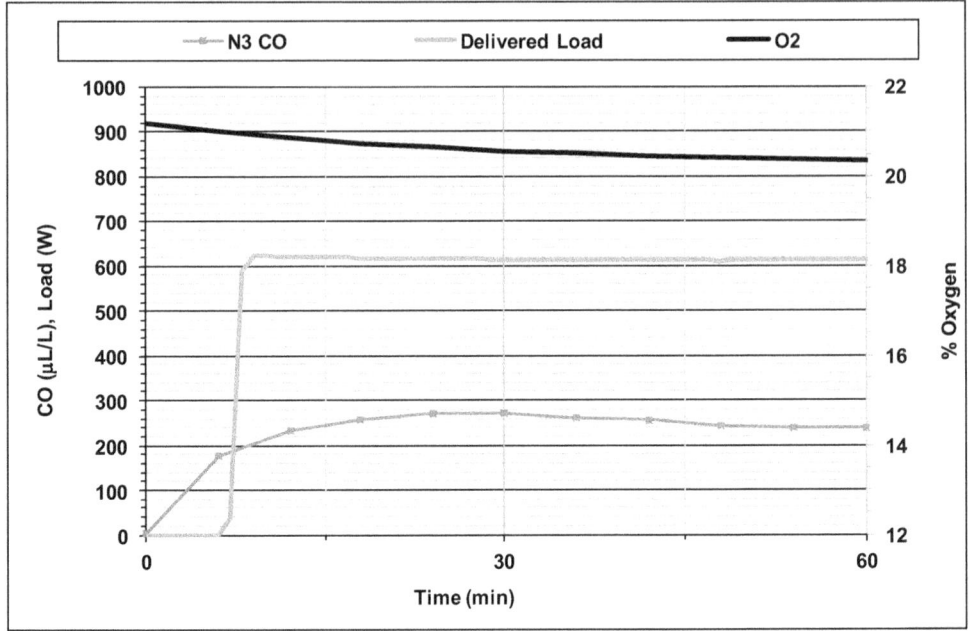

Figure C7b CO concentrations in the house for Test AV (Gen SO1, Configuration 1)

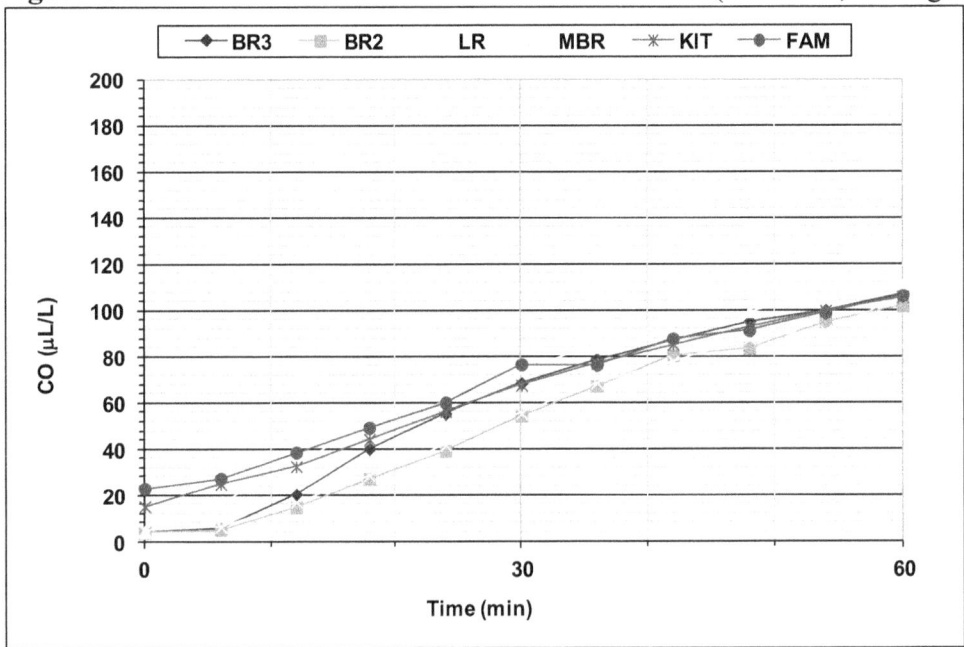

Figures C8a and C8b show the results for Test AI, which was a 1.1 h test of Gen B in Configuration 1 with a constant 5500 W load. The concentration of CO in the garage reached a peak of about 5500 μL/L and the volume fraction of O_2 in the garage dropped to about 19.6 % when the generator was stopped.

Figure C8a CO and O_2 concentrations in the garage and measured load for Test AI (Gen B, Configuration 1)

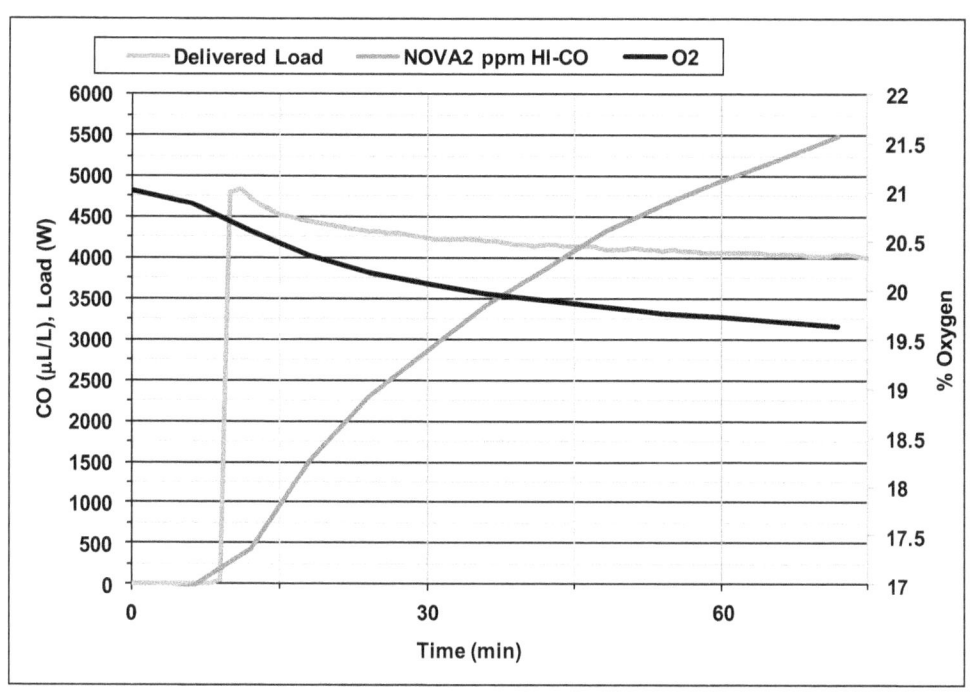

Figure C8b CO concentrations in the house for Test AI (Gen B, Configuration 1)

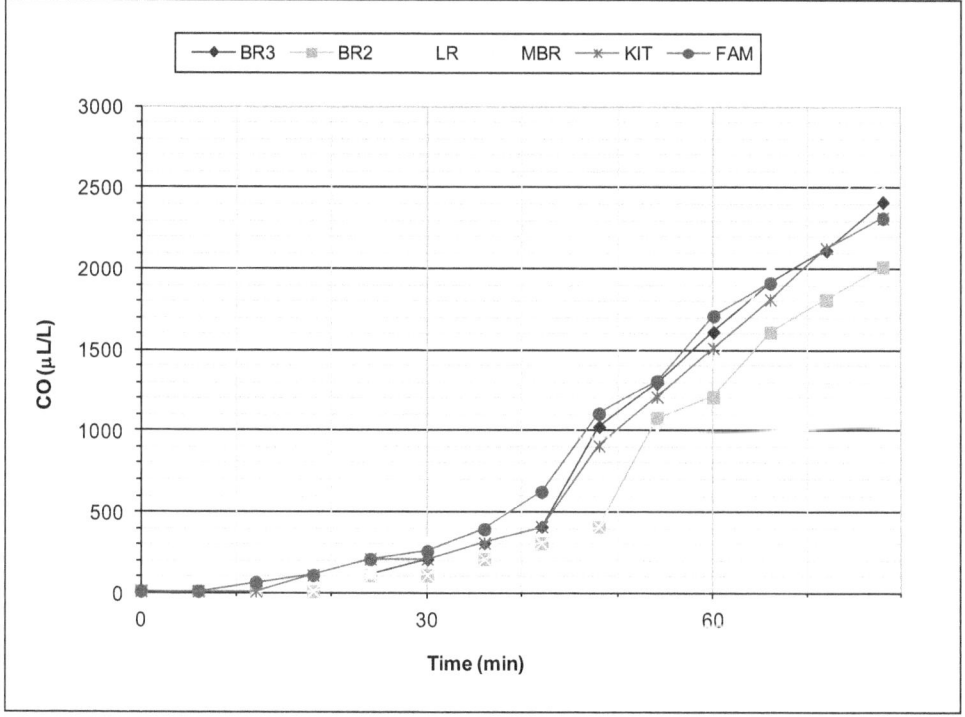

Figures C9a and C9b show the results for Test AR, which was a 4 h test of Gen SO1 with test house configuration 1 and a constant 5500 W load. During this test, the load bank overheated multiple times (resulting in the temporary sharp drops in load seen in Figure 9a) and, following the test, the exhaust manifold was found to be loose, which may have impacted the test results.

Figure C9a CO and O_2 concentrations in the garage and measured load for Test AR (Gen SO1, Configuration 1)

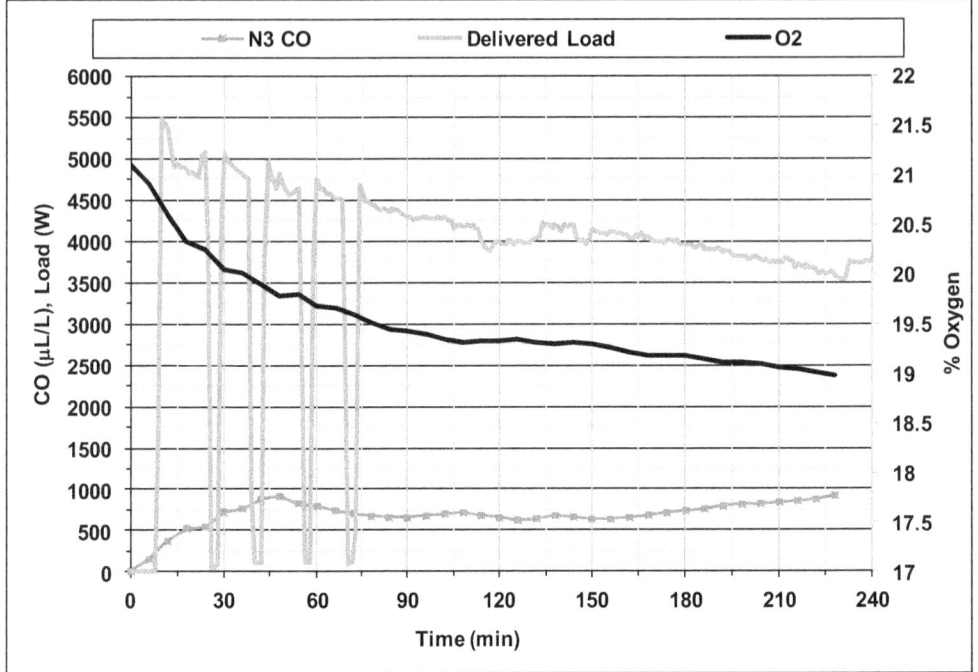

Figure C9b CO concentrations in the house for Test AR (Gen SO1, Configuration 1)

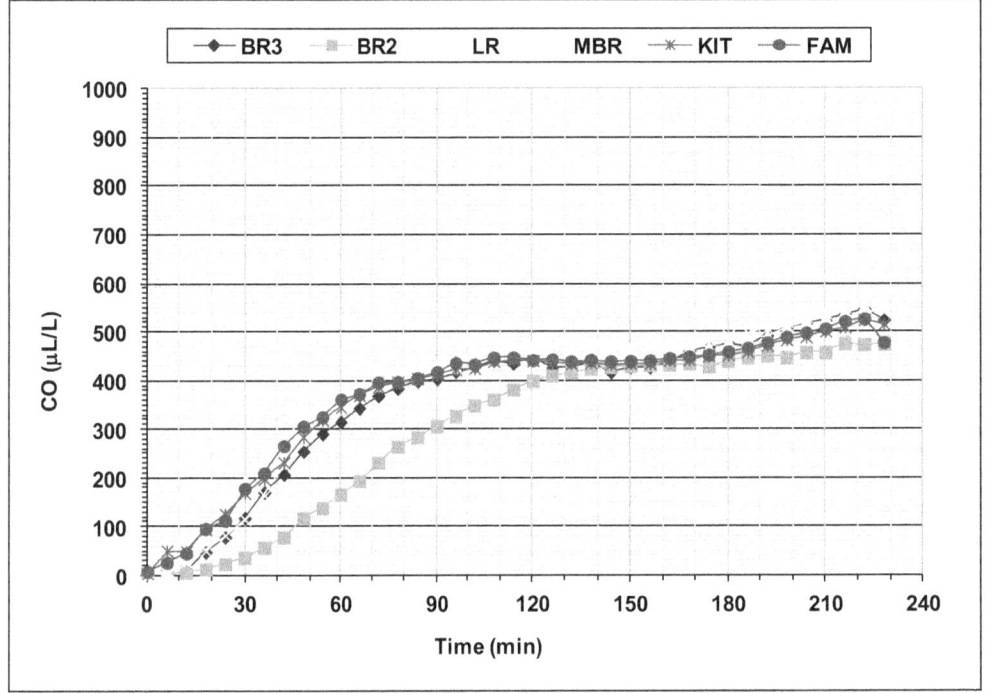

Figures C10a and C10b show the results for Test AS, which was also a 4 h test of Gen SO1 with test house configuration 1 and a constant 5500 W load. As with Test AR, the load bank overheated multiple times during this test. The results of Test AR were very similar to those of Test AS.

Figure C10a CO and O_2 concentrations in the garage and measured load for Test AS (Gen SO1, Configuration 1)

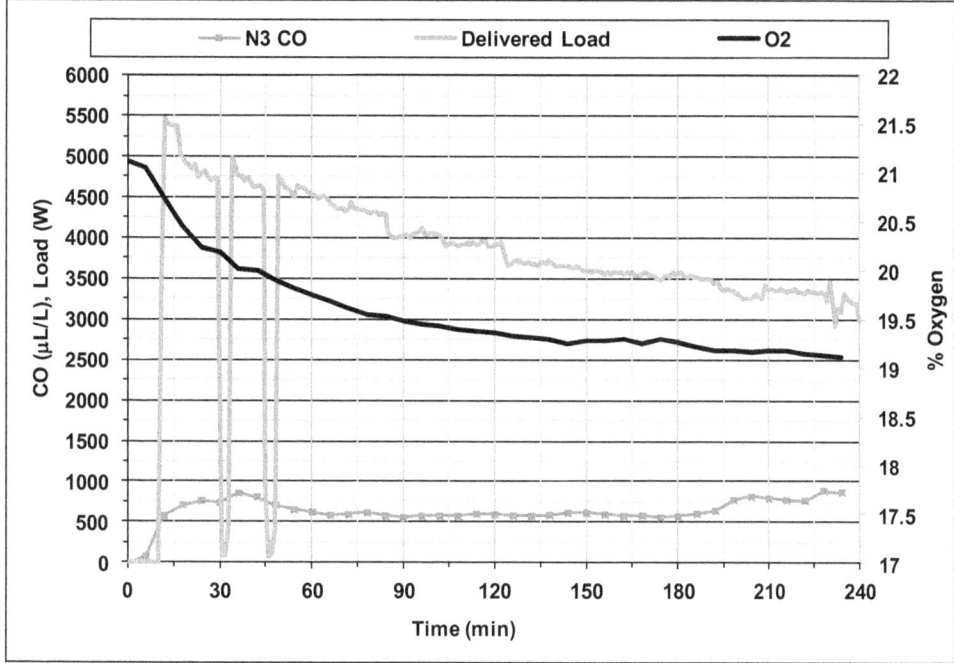

Figure C10b CO concentrations in the house for Test AS (Gen SO1, Configuration 1)

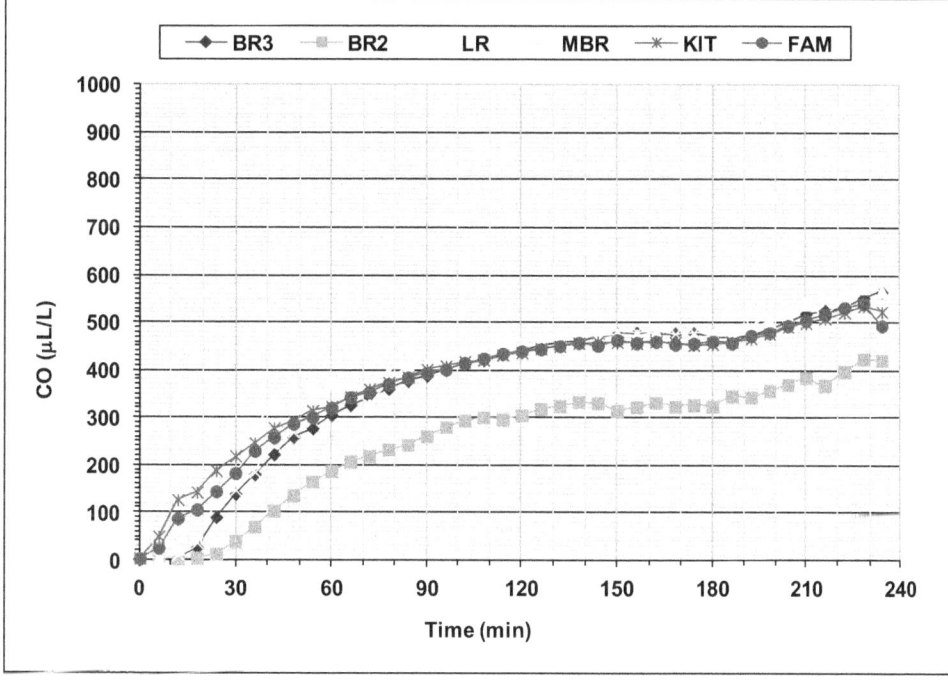

Figures C11a and C11b show the results for Test AL, which was a test of Gen SO1 with the shutoff algorithm enabled with test house configuration 1 and a constant 5500 W load. The algorithm did not shut off the generator before it was manually stopped and the test terminated after about 40 min when a generator circuit breaker tripped.

Figure C11a CO and O_2 concentrations in the garage and measured load for Test AL (Gen SO1, Configuration 1)

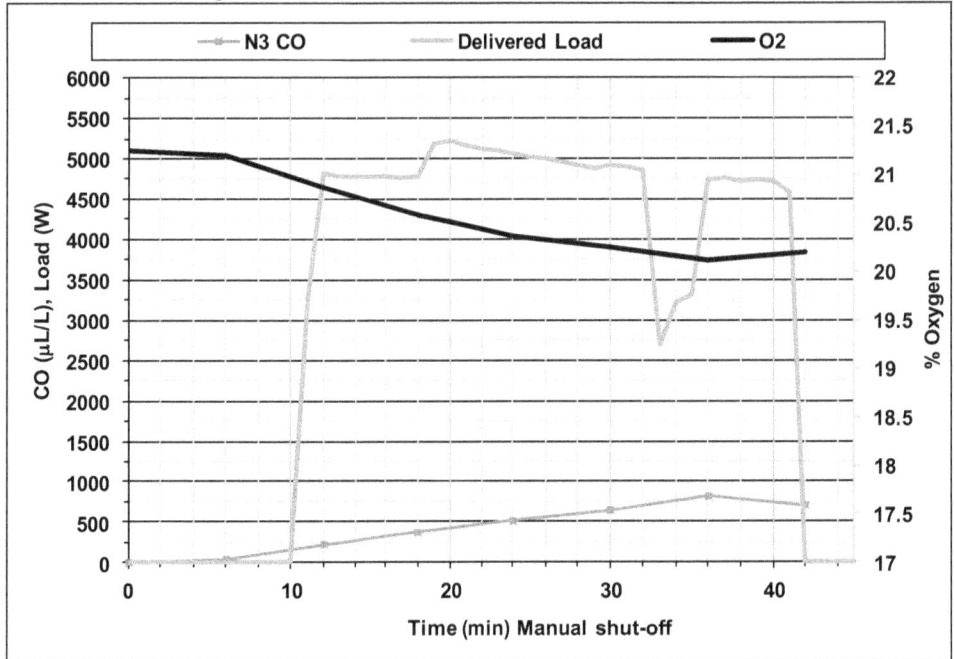

Figure C11b CO concentrations in the house for Test AL (Gen SO1, Configuration 1)

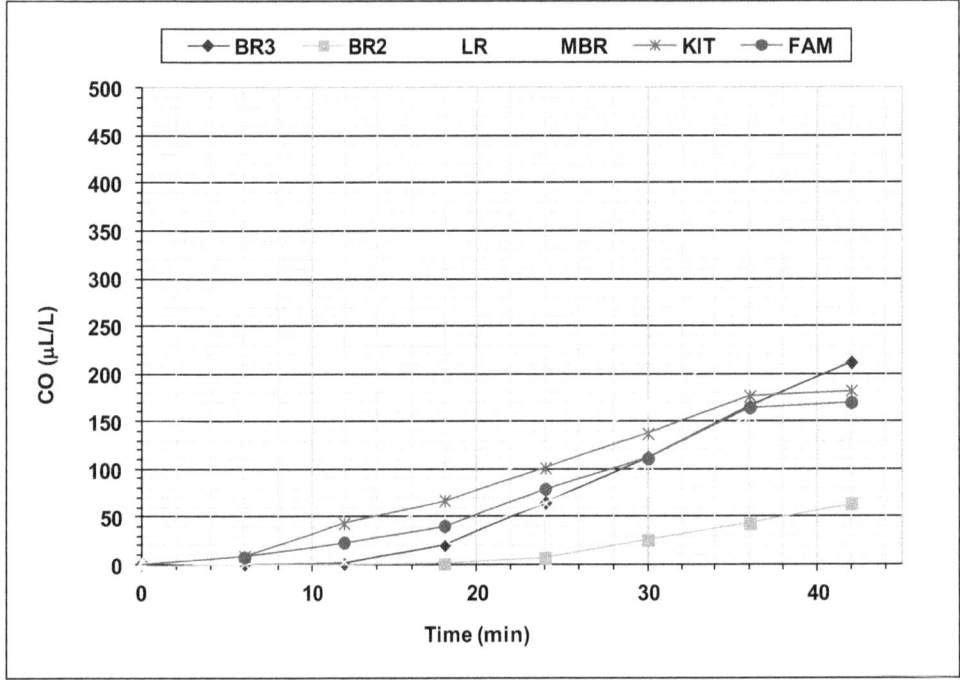

Figures C12a and C12b show the results for Test AM, which was also a test of Gen SO1 with the shutoff algorithm enabled under test house configuration 1 and a constant 5500 W load. The algorithm shut off the generator after about 7 min, which was followed by a 2 h natural decay period before mechanical venting of the house.

Figure C12a CO and O_2 concentrations in the garage and measured load for Test AM (Gen SO1, Configuration 1)

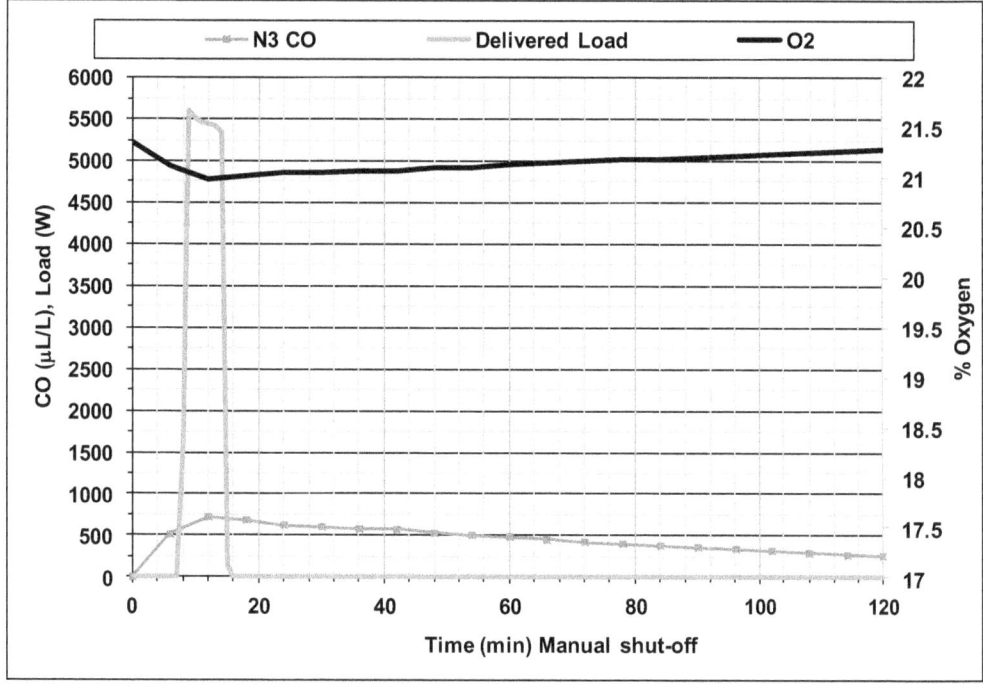

Figure C12b CO concentrations in the house for Test AM (Gen SO1, Configuration 1)

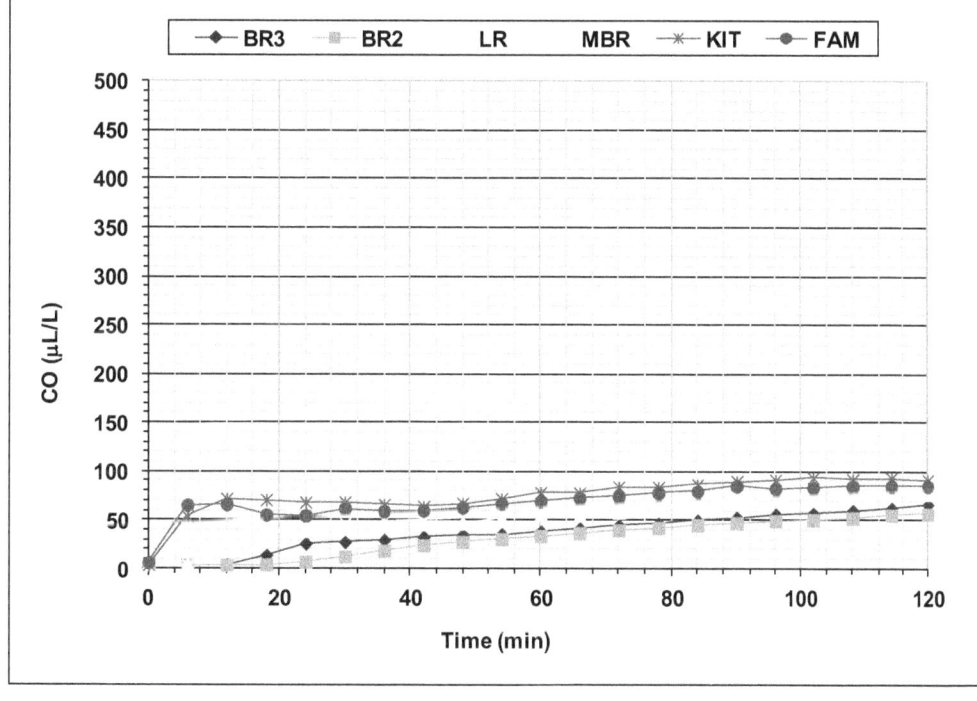

Figures C13a and C13b show the results for Test AN, which was a test of Gen SO1 with the shutoff algorithm enabled under test house configuration 1 and a constant 2500 W

121

load. The algorithm shut off the generator after about 13 min which was followed by a 1 h natural decay period before mechanical venting of the house.

Figure C13a CO and O_2 concentrations in the garage and measured load for Test AN (Gen SO1, Configuration 1)

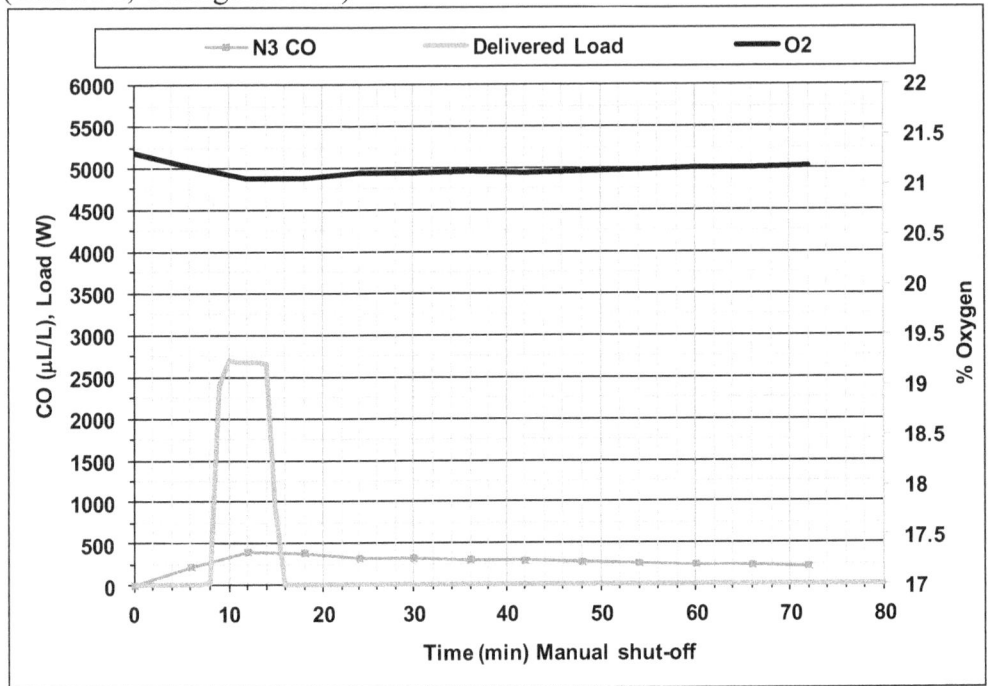

Figure C13b CO concentrations in the house for Test AN (Gen SO1, Configuration 1)

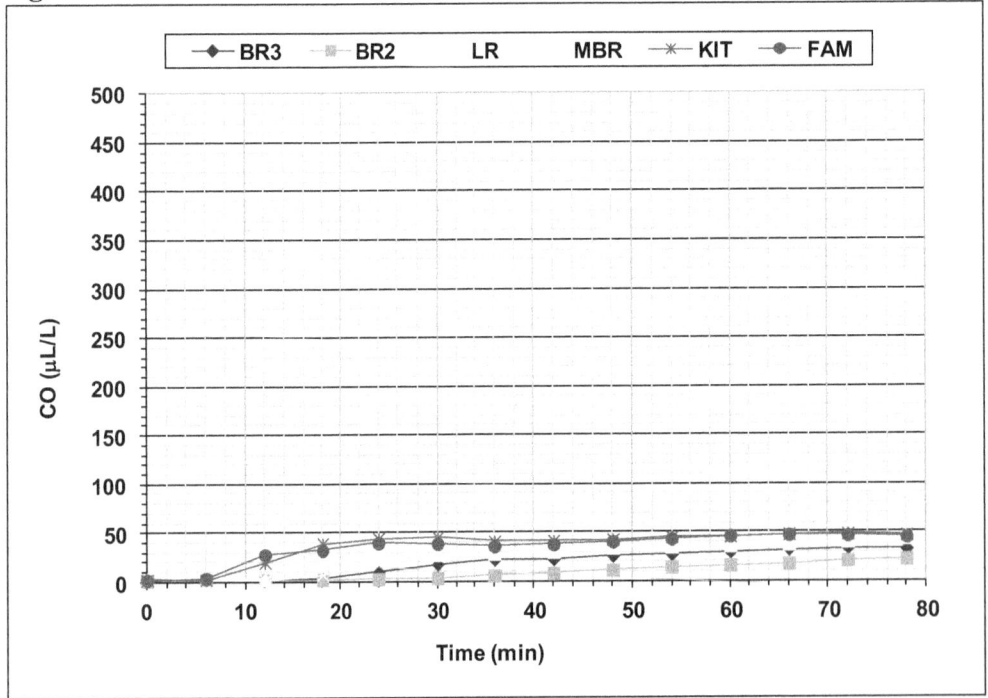

Figures C14a and C14b show the results for Test H, which was a 4 h test of unmod Gen X with Configuration 2 (garage bay door open, garage access door to house closed, and the house central HVAC fan off). The results of Test H were similar to those for Test F which was the same generator and house configuration. The peak concentration occurred during the 1500 W setting and rose slightly in each load cycle, reaching a maximum concentration near 1200 µL/L in the fourth load cycle. For this test, the garage was not instrumented with a low concentration CO analyzer, and the instrument uncertainty is large relative to measured concentrations below 500 µL/L. During the course of this test, with the garage bay door open, the oxygen level dipped only slightly, and the delivered electrical output was consistent during each cycle, largely meeting the load bank setting with the exception of the 5500 W setting.

Figure C14a CO and O_2 concentrations in the garage and measured load for Test H (unmod Gen X, Configuration 2)

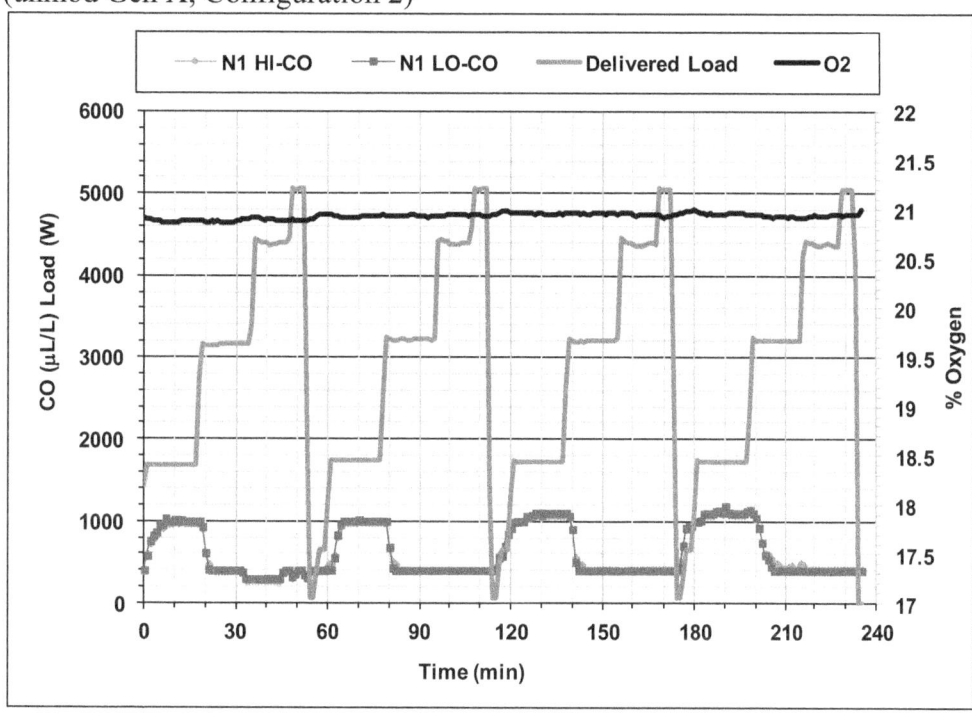

Figure C14b CO concentrations in the house for Test H (unmod Gen X, Configuration 2)

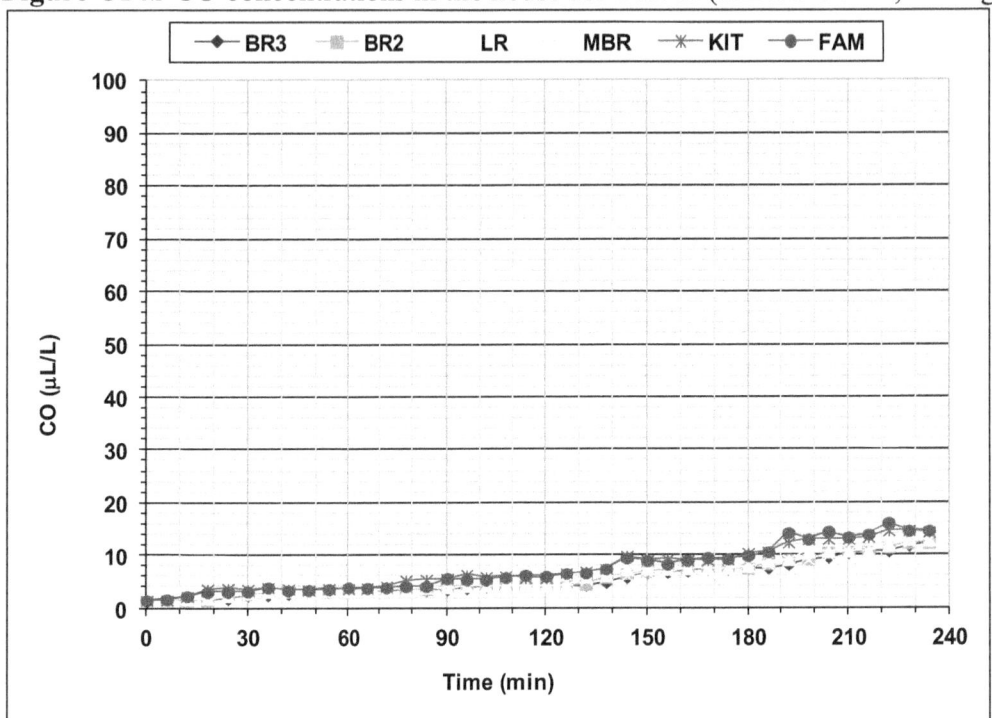

Figures C15a and C15b show the results for Test S, which was a shut-off test of Gen SO1with the shutoff algorithm enabled under test house configuration 2 and the cyclic load. The algorithm shut off the generator after approximately 0.9 h. The results shown in Figure C15 for Test S, prior to generator shut-off, are similar to those in Figure 18 in the body of the report for Test T (a test of the same generator and house configuration but without the shut-off activated).

Figure C15a CO and O_2 concentrations in the garage and measured load for Test S (Gen SO1, Configuration 2)

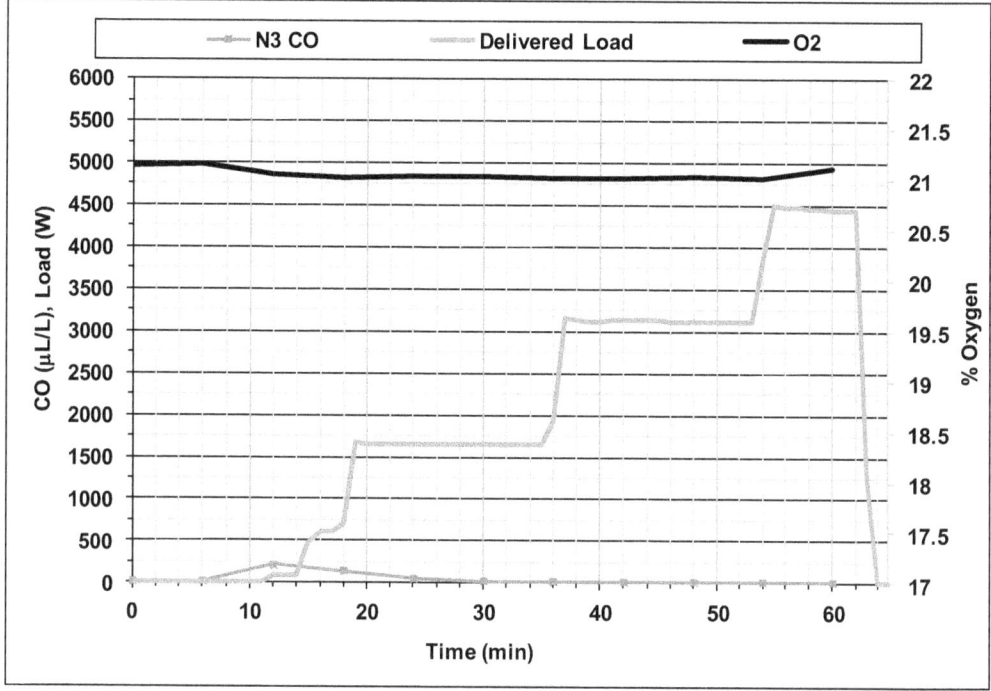

Figure C15b CO concentrations in the house for Test S (Gen SO1, Configuration 2)

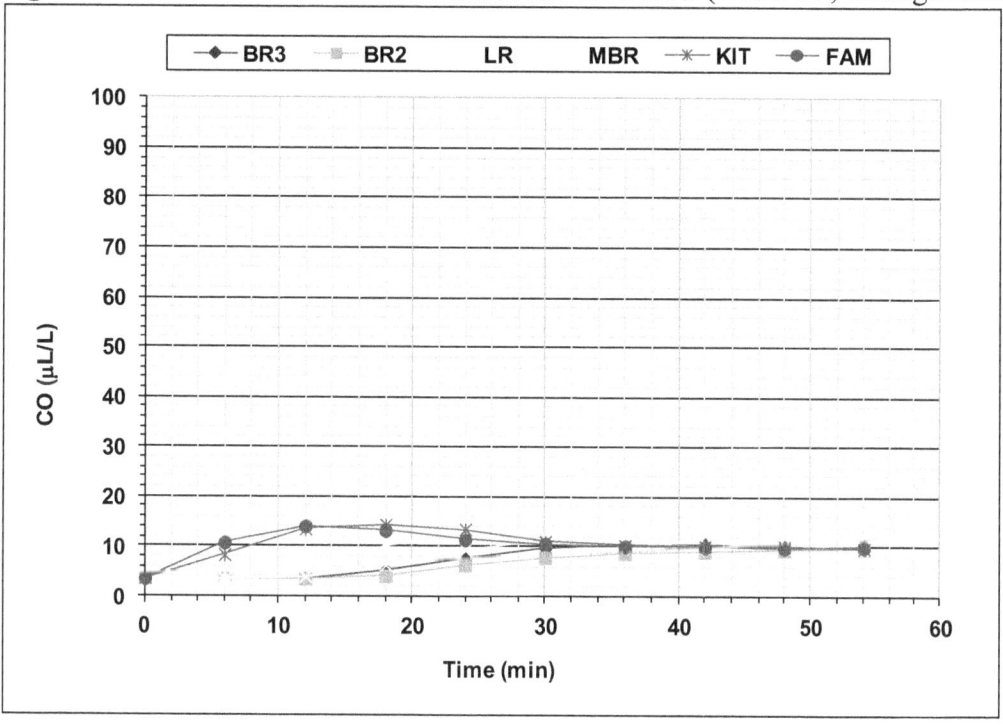

Figures C16a and C16b show the results for Test Q, which was a 4 h test of Gen B with the house in Configuration 2 under the cyclic load. Figure C16a shows the concentration of CO in the garage reached a peak of near 500 μL/L and the volume fraction of O$_2$ in the garage did not measurably change when the generator was stopped.

Figure C16a CO and O$_2$ concentrations in the garage and measured load for Test Q (Gen B, Configuration 2)

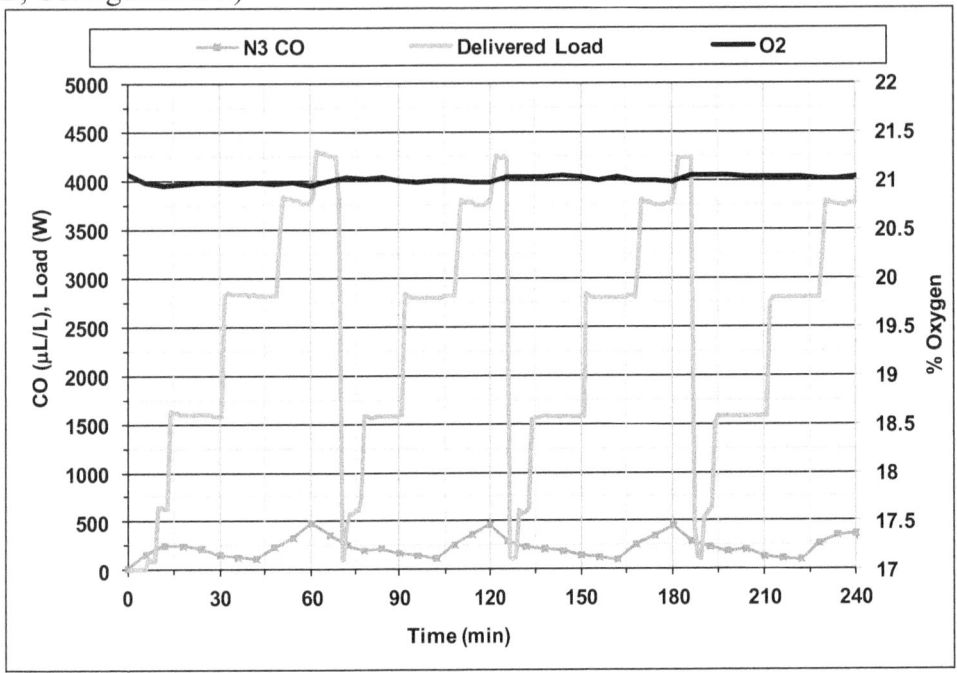

Figure C16b CO (ppm range) concentrations in the house for Test Q (Gen B, Configuration 2)

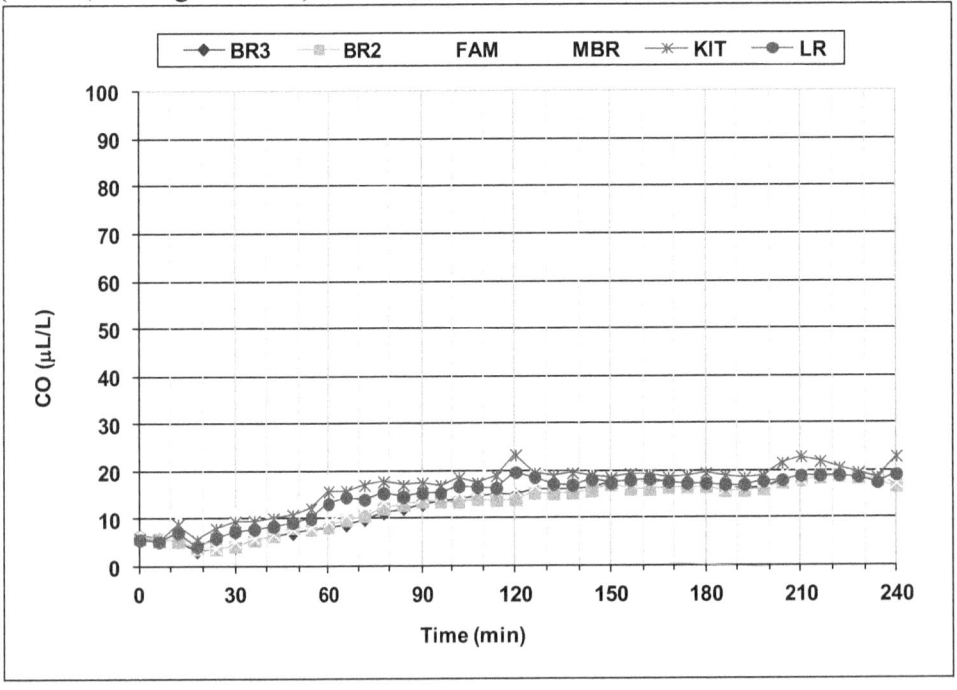

Figures C17a and C17b show the results for Test A, which was a 1.5 h test of unmod Gen X in Configuration 3 (garage bay door closed, garage access door to house open two inches, and the house central HVAC fan on). These conditions are the same as Test I and the results of Test A were similar to the first 1.5 h of Test I (see Figure 19 in the body of the report). After the generator was manually stopped, the garage and house were mechanically vented. Figure C17a shows that the concentration of CO in the garage reached a peak of about 9200 μL/L and the concentration of O_2 in the garage dropped to 18.6 % when the generator was stopped. It also shows that in the first load cycle the delivered electrical output was less than the load bank settings for the two highest loads in the load cycle, 4500 W and 5500 W, which were applied as the oxygen was approaching 19 %.

Figure C17a CO and O_2 concentrations in the garage and measured load for Test A (unmod Gen X, Configuration 3)

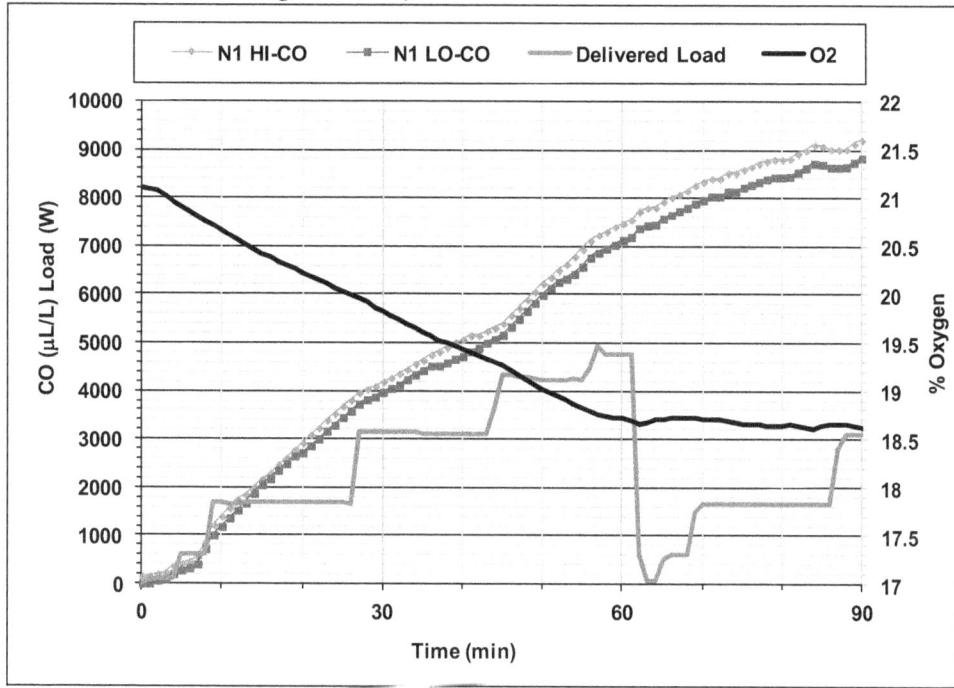

Figure C17b CO (high range) concentrations in the house for Test A (unmod Gen X, Configuration 3)

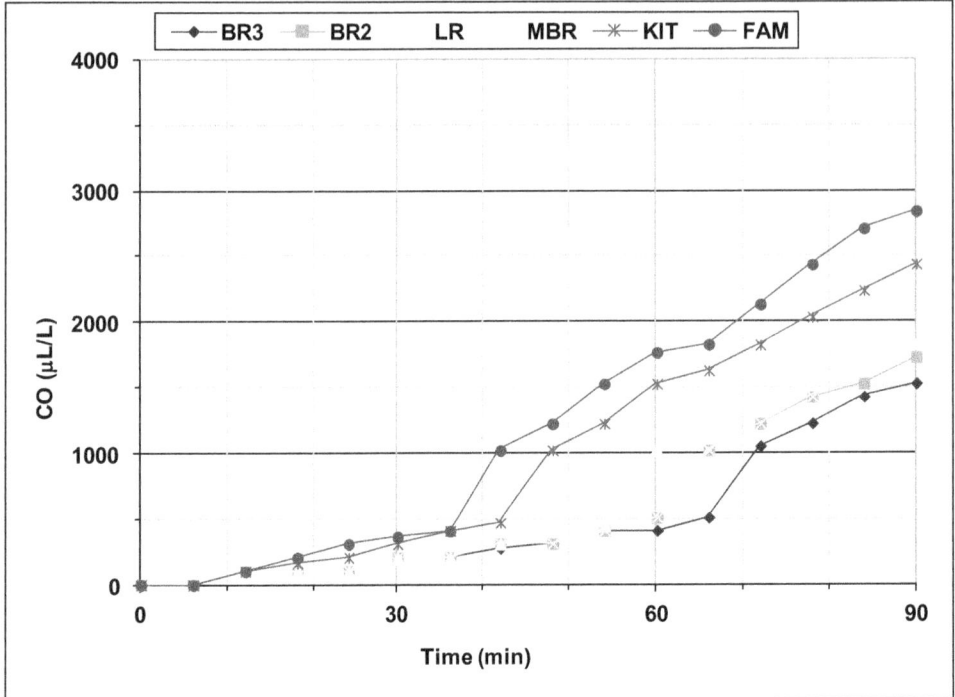

Figures C18a and C18b show the results for Test C, which was a 4 h test of unmod Gen X in Configuration 3 followed by a 1 h natural decay. These conditions are the same as Test I and the results of Test C were similar to those of Test I (see Figure 19 in the body of the report). Figure C18a shows that the concentration of CO in the garage reached a peak of about 21,000 μL/L and the concentration of O_2 in the garage dropped to 17.0 % when the generator was stopped. It also shows that during the first load cycle the delivered electrical output was less than the load bank settings for the two highest loads in the cycle, 4500 W and 5500 W, which were applied as the oxygen was around 19 %. As the oxygen continued to drop in the subsequent load cycles, the delivered power for these two load points decreased further.

Figure C18a CO and O_2 concentrations in the garage and measured load for Test C (unmod Gen X, Configuration 3)

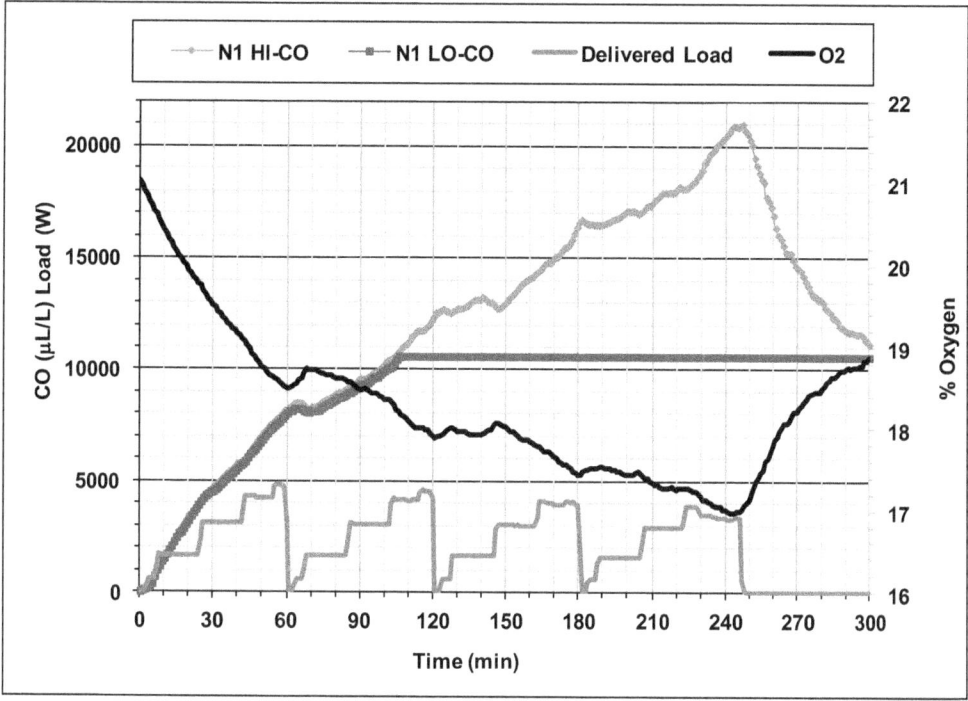

Figure C18b CO (high range) concentrations in the house for Test C (unmod Gen X, Configuration 3)

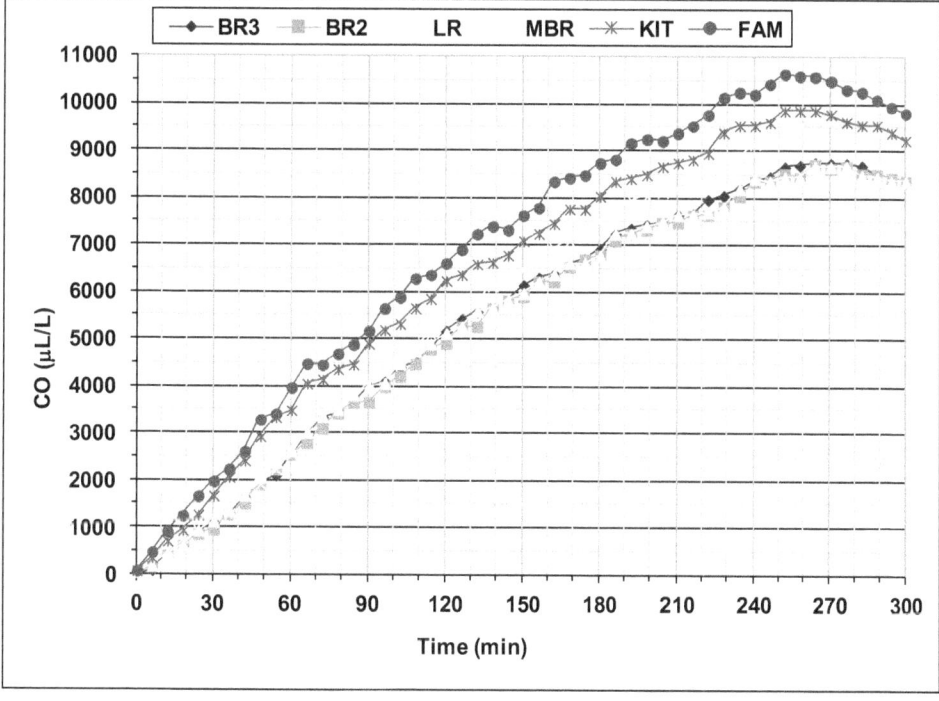

Figures C19a and C19b show the results for Test AA, which was a shut-off test of Gen SO1 with the shutoff algorithm enabled with test house configuration 3 and the cyclic load. The algorithm shut off the generator after approximately 1.6 h. A natural decay period of about 1 h was included after the generator was stopped, followed by mechanical venting. The results shown in Figure C19 for Test AA, prior to generator shut-off, are similar to those in Figure 20 (in the body of the report) for Test Z (a test of the same generator and house configuration but without the shut-off activated) although the CO concentration reached a higher level during the initial spike for Test AA.

Figure C19a CO and O_2 concentrations in the garage and measured load for Test AA (Gen SO1, Configuration 3)

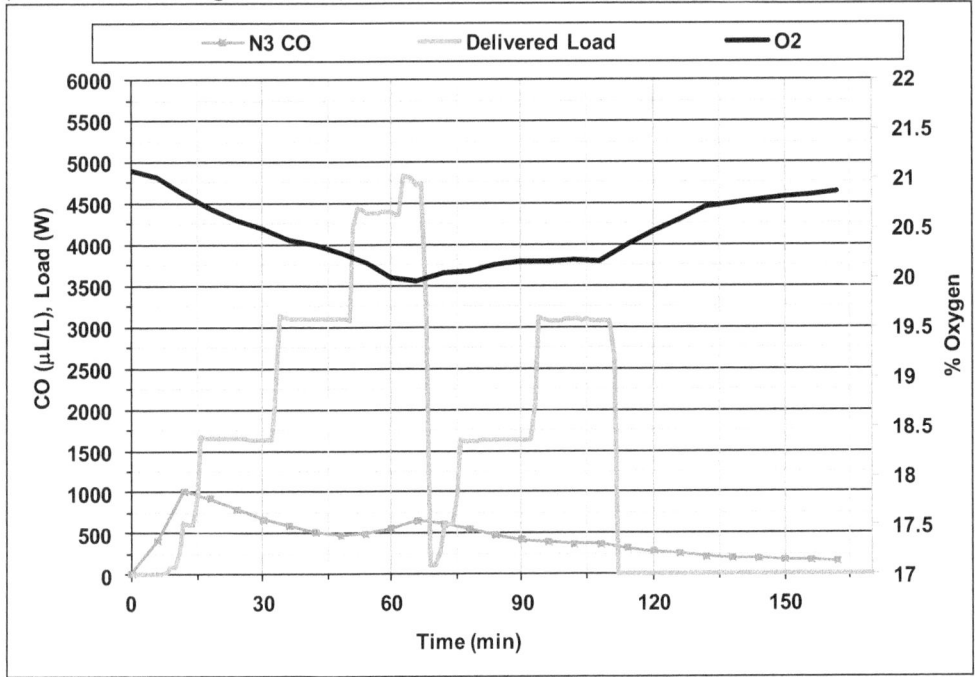

Figure C19b CO concentrations in the house for Test AA (Gen SO1, Configuration 3)

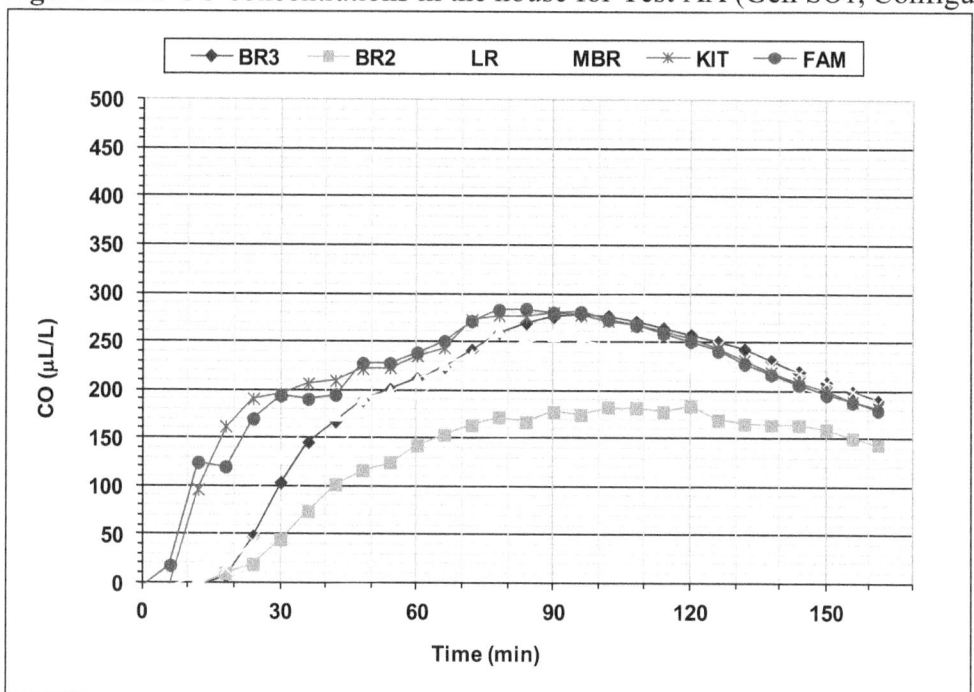

Figures C20a and C20b show the results for Test X, which was a shut-off test of Gen SO1 with the shutoff algorithm enabled under test house configuration 4 and the cyclic load. The algorithm shut off the generator after approximately 0.8 h. A natural decay period of about 1.2 h was included after the generator was stopped, followed by mechanical venting. The results shown in Figure C20 for Test X, prior to generator shut-off, are similar to those in Figure 20 for Test Z (a test of the same generator and house configuration but without the shut-off activated).

Figure C20a CO and O₂ concentrations in the garage and measured load for Test X (Gen SO1, Configuration 4)

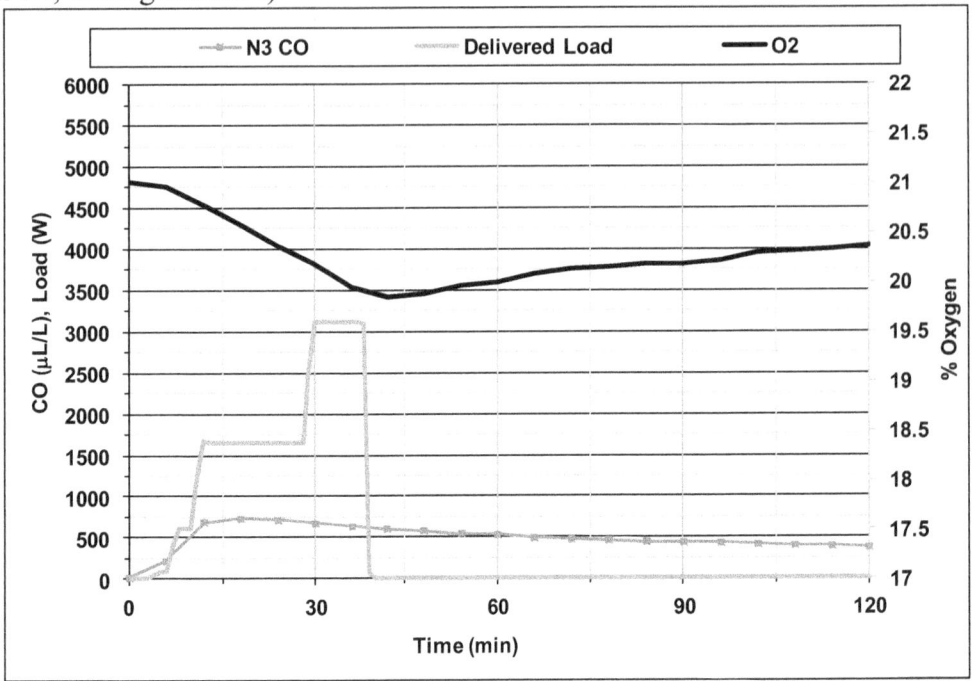

Figure C20b CO concentrations in the house for Test X (Gen SO1, Configuration 4)

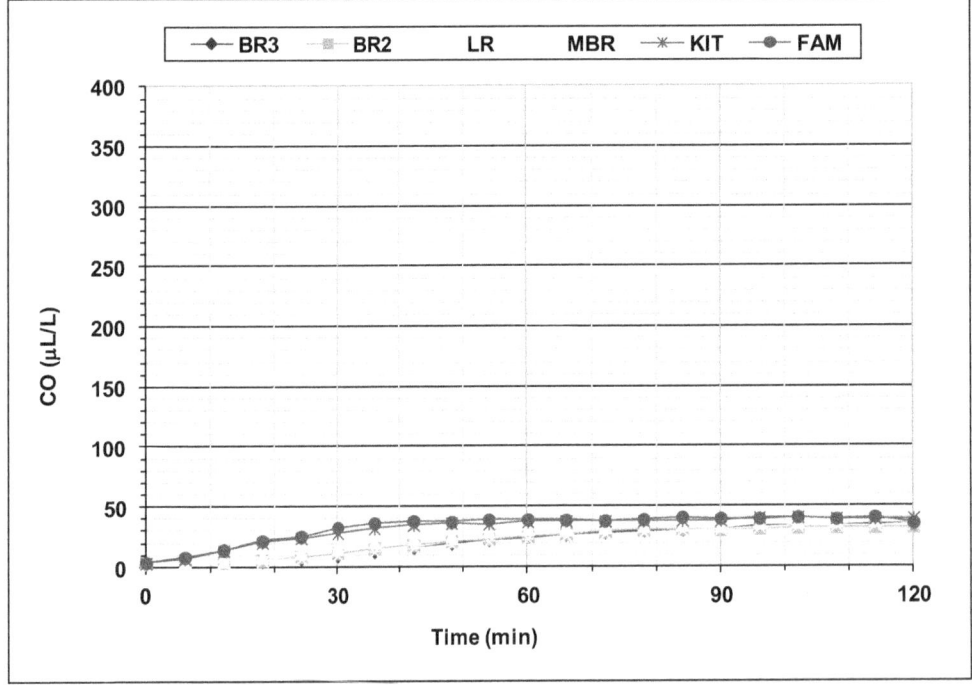

Figures C21a and C21b show the results for Test AB, which was a shut-off test of Gen SO1 with the shutoff algorithm enabled under test house configuration 5 and the cyclic load. The algorithm shut off the generator after approximately 1 h. A natural decay period of about 1.1 h was included after the generator was stopped, followed by mechanical venting. The results shown in Figure C21 for Test AB, prior to generator shut-off, are similar to those in Figure 24 (in the body of the report) for Test AH (a test of the same generator and house configuration but without the shut-off activated) although the initial spike in CO concentration in the garage was larger for Test AB.

Figure C21a CO and O_2 concentrations in the garage and measured load for Test AB (Gen SO1, Configuration 5)

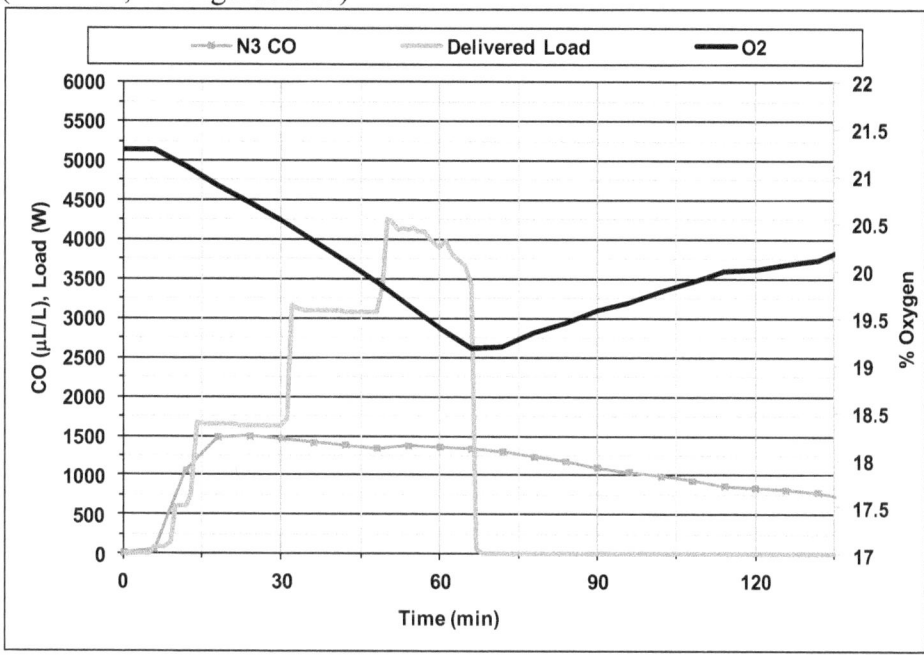

Figure C21b CO concentrations in the house for Test AB (Gen SO1, Configuration 5)

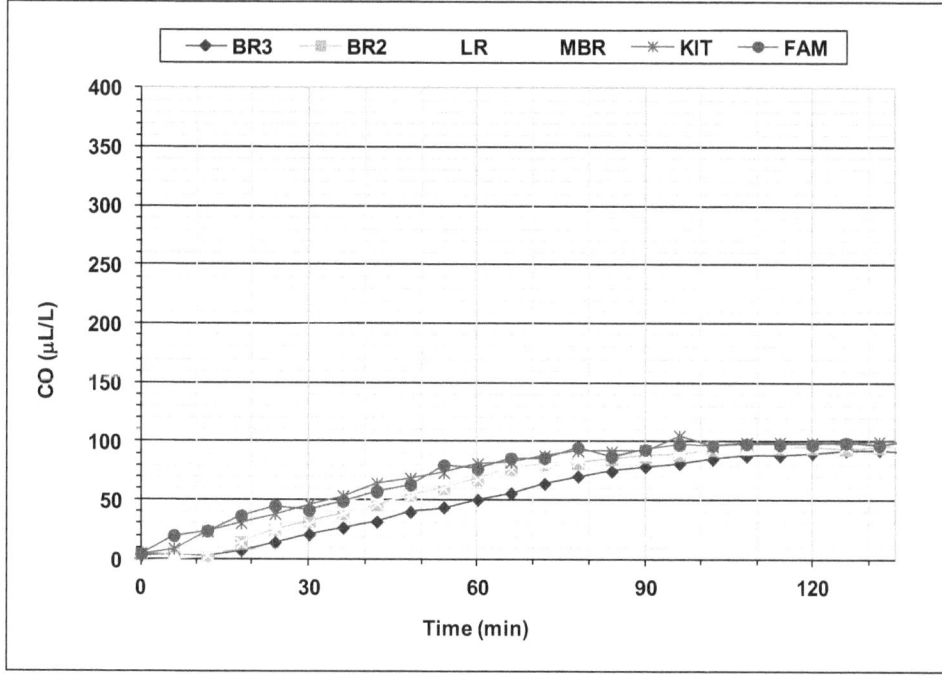

Figures C22a and C22b show the results for Test AQ, which was also a shut-off test of Gen SO1with the shutoff algorithm enabled under test house configuration 5 and the cyclic load. The algorithm shut off the generator after approximately 0.8 h. A natural decay period of about 1.5 h was included after the generator was stopped, followed by mechanical venting. Figure C22a shows no initial spike in CO concentration in the garage for Test AQ because Test AQ was a warm start while Test AB was a cold start.

Figure C22a CO and O_2 concentrations in the garage and measured load for Test AQ (Gen SO1, Configuration 5)

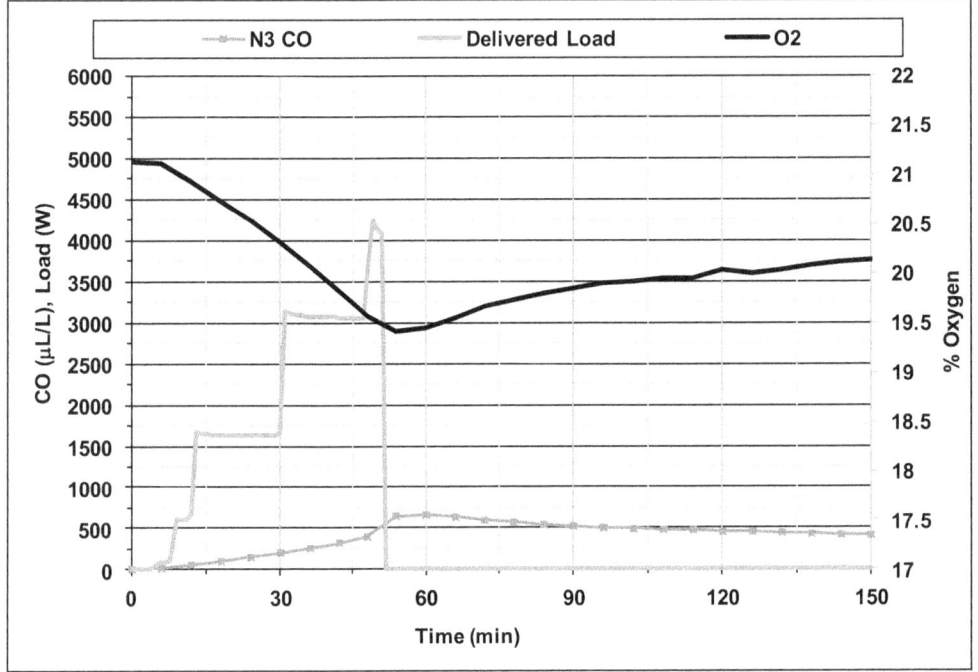

Figure C22b CO concentrations in the house for Test AQ (Gen SO1, Configuration 5)

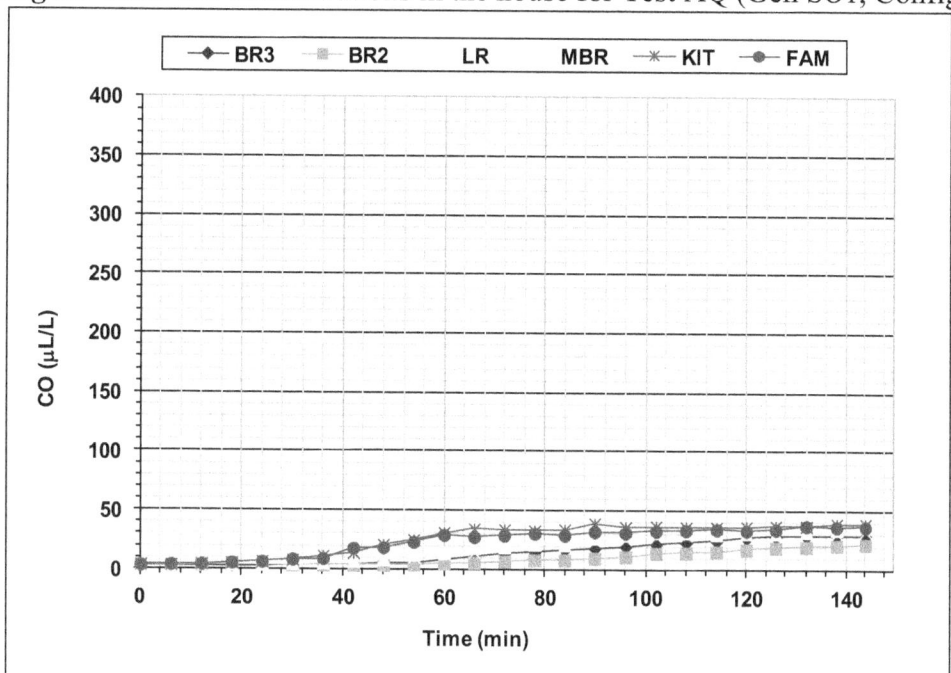

Figures C23a and C23b show the results for Test Y, which was a shut-off test of Gen SO1with the shutoff algorithm enabled under test house configuration 6 and the cyclic load. The algorithm shut off the generator after approximately 0.5 h, followed by mechanical venting. The results shown in Figure C23 for Test Y, prior to generator shut-off, are very similar to those in Figure 26 (in the body of the report) for Test U (a test of the same generator and house configuration but without the shut-off activated).

Figure C23a CO and O_2 concentrations in the garage and measured load for Test Y (Gen SO1, Configuration 6)

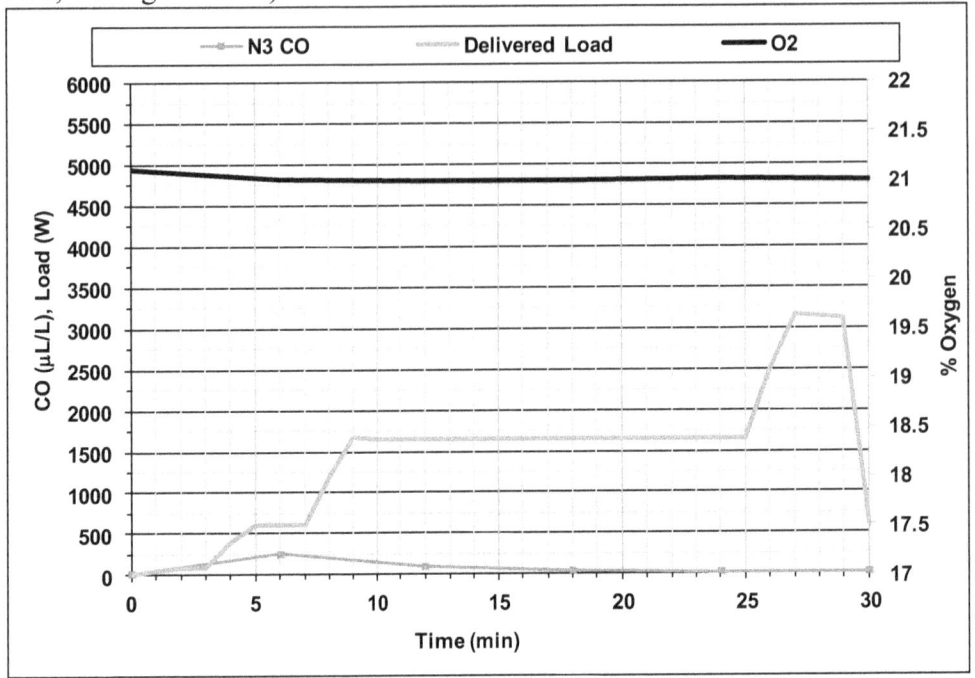

Figure C23b CO concentrations in the house for Test Y (Gen SO1, Configuration 6)

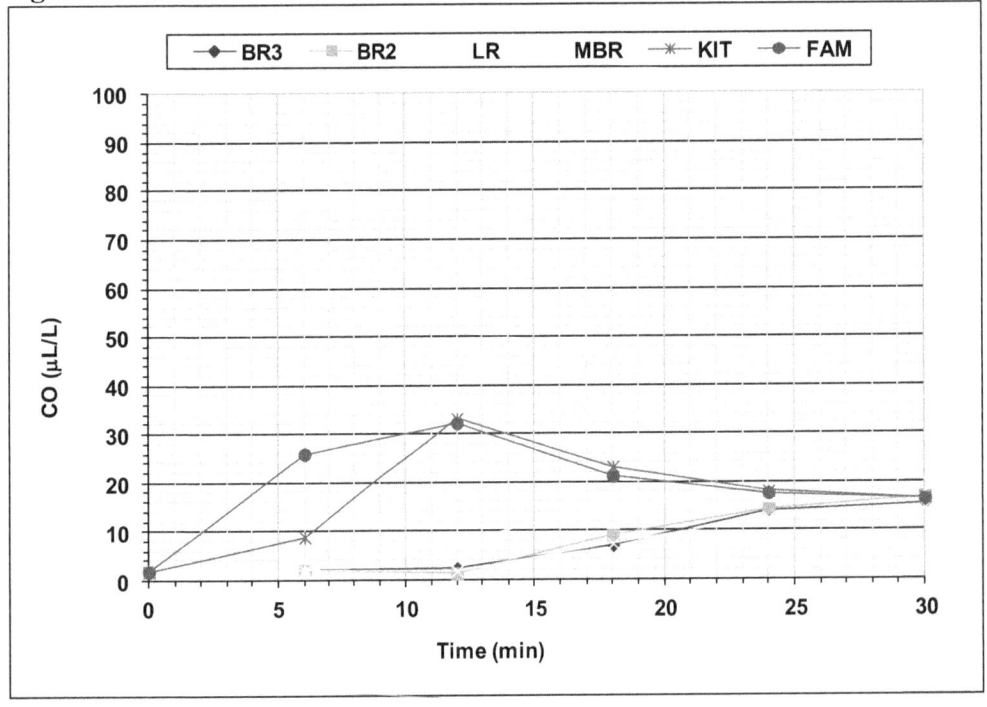

Figures C24a and C24b show the results for Test AO, which was a shut-off test of Gen SO1with the shutoff algorithm enabled under test house configuration 7 and the cyclic load. The algorithm shut off the generator after approximately 1.1 h, followed by mechanical venting. The results shown in Figure C24 for Test AO do not show an initial spike in garage CO concentration like that in Figure 28 (in the body of the report) for Test V (a test of the same generator and house configuration but without the shut-off activated) but with a somewhat smaller initial spike in concentration.

Figure C24a CO and O_2 concentrations in the garage and measured load for Test AW (Gen SO1, Configuration 7)

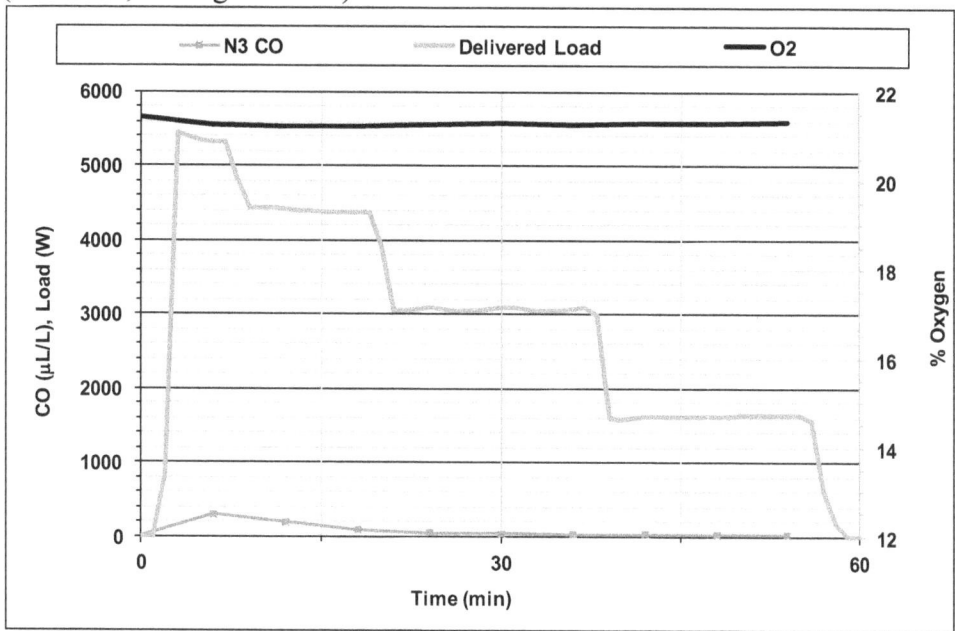

Figure C24b CO concentrations in the house for Test AW (Gen SO1, Configuration 7)

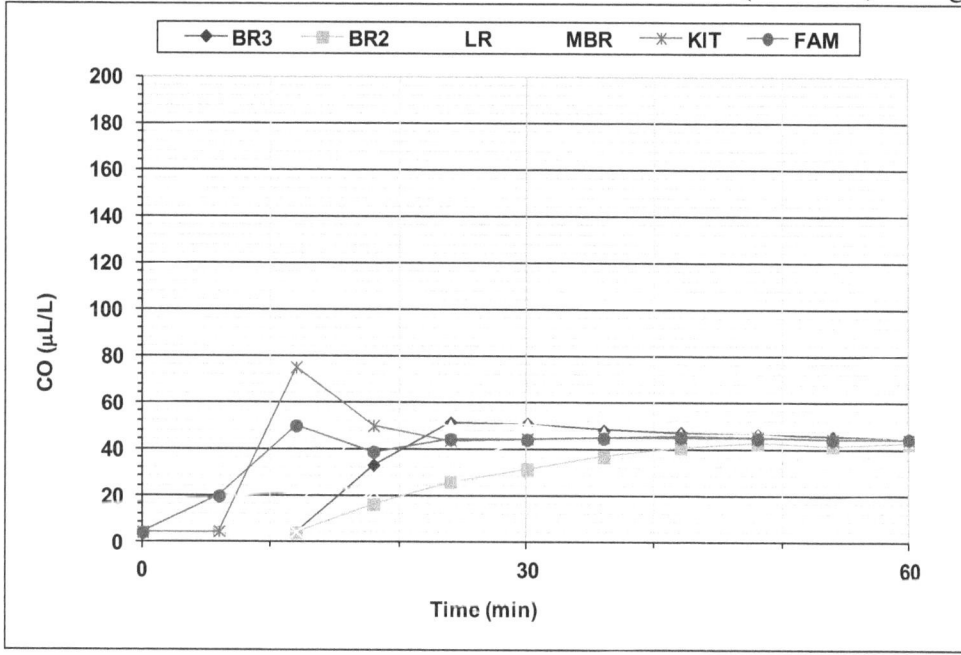

Figures C25a and C25b show the results for Test AO, which was a shut-off test of Gen SO1 with the shutoff algorithm enabled under test house configuration 7 and the cyclic load. The algorithm shut off the generator after approximately 1.1 h, followed by mechanical venting. The results shown in Figure C25 for Test AO do not show an initial spike in garage CO concentration like that in Figure C24 for Test AW (a test of the same generator and house configuration but loaded in a high to low cyclic pattern), likely because Test AO was a warm start.

Figure C25a CO and O₂ concentrations in the garage and measured load for Test AO (Gen SO1, Configuration 7)

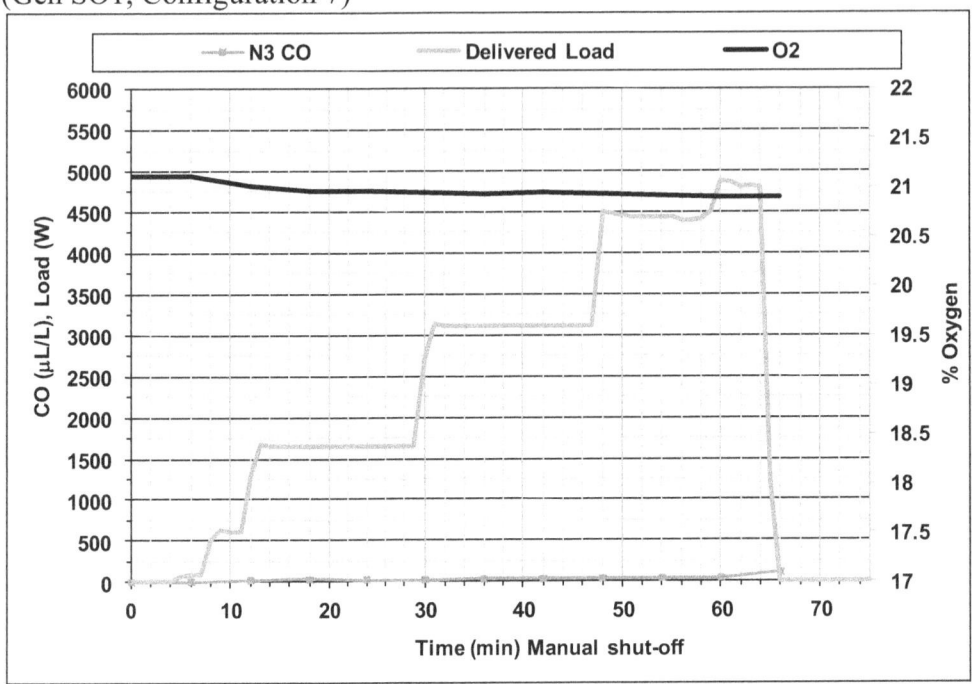

Figure C25b CO concentrations in the house for Test AO (Gen SO1, Configuration 7)

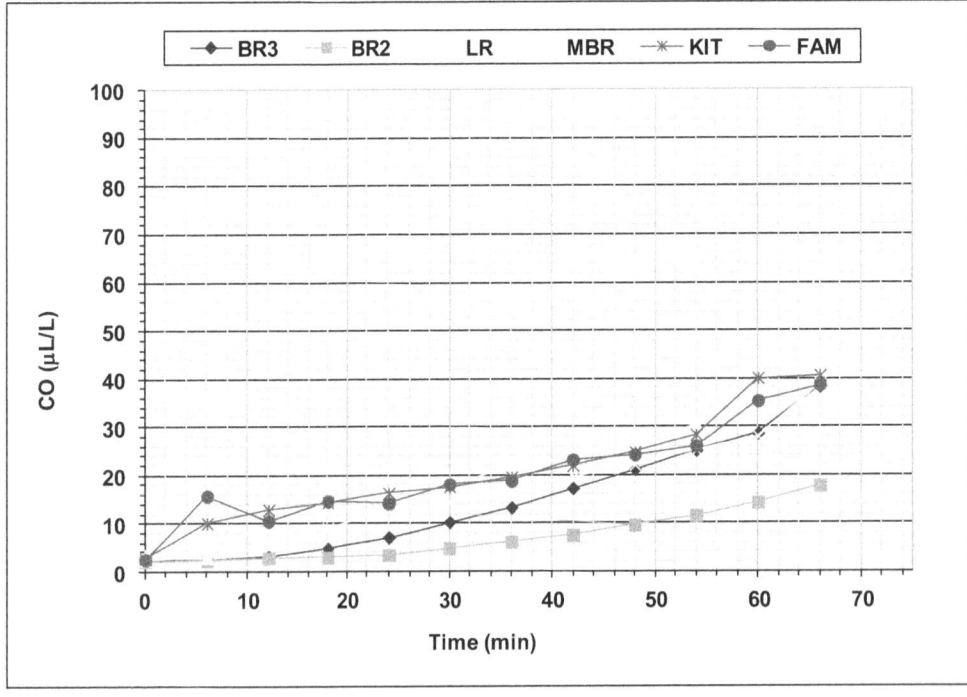

Figures C26a and C26b show the results for Test AJ, which was a 3 h test of Gen B with the house in Configuration 8 with a 5500 W load. Figure C26a shows the concentration of CO in the garage reached a peak near 200 μL/L and the volume fraction of O_2 in the garage did not measurably change during the test.

Figure C26a CO and O_2 concentrations in the garage and measured load for Test AJ (Gen B, Configuration 8)

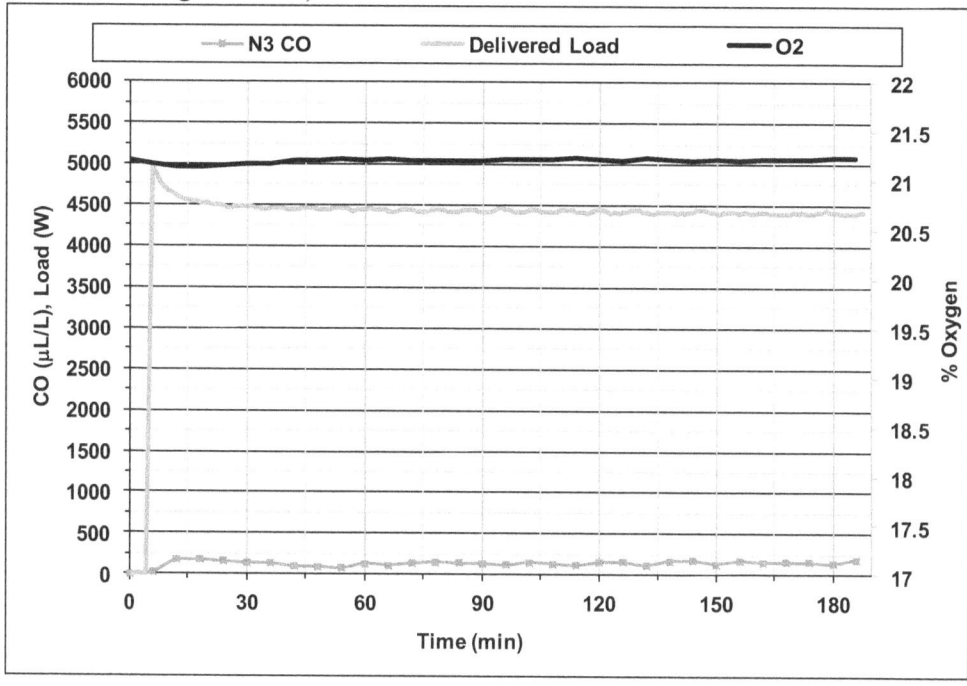

Figure C26b CO concentrations in the house for Test AJ (Gen B, Configuration 8)

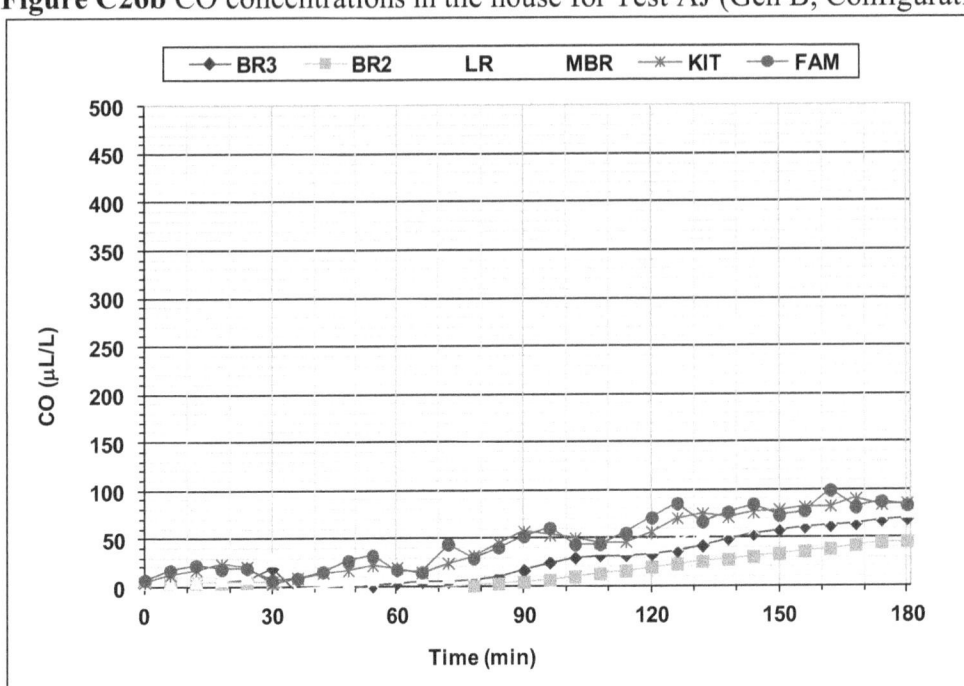

Figures C27a and C27b show the results for Test AK, which was a shut-off test of Gen SO1with the shutoff algorithm enabled under test house configuration 8 and a 5500 W load. The algorithm shut off the generator after approximately 1.4 h, followed by mechanical venting.

Figure C27a CO and O_2 concentrations in the garage and measured load for Test AK (Gen SO1, Configuration 8)

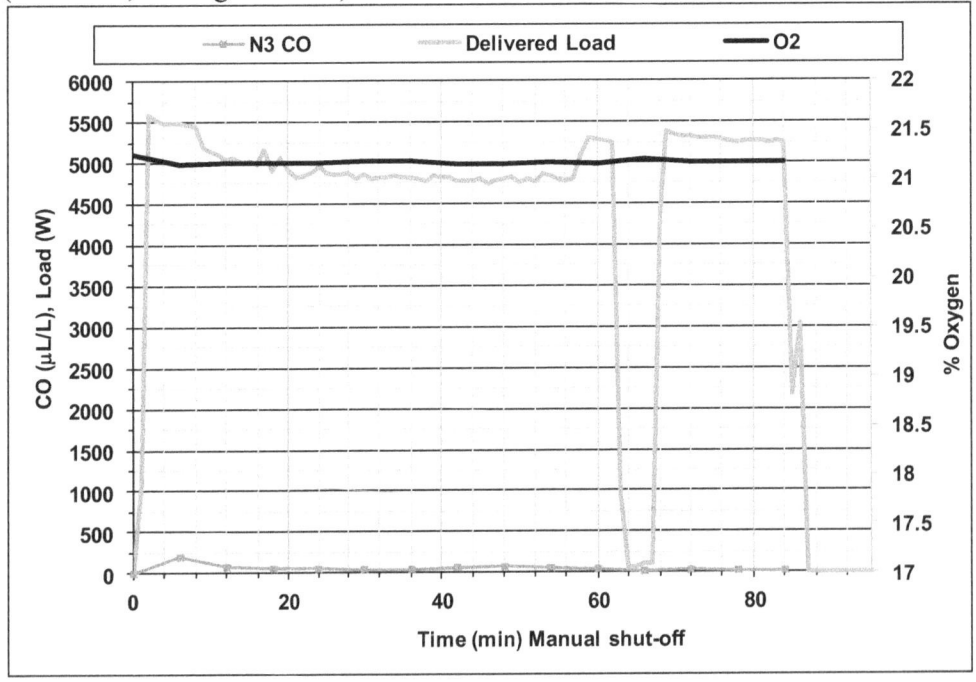

Figure C27b CO concentrations in the house for Test AK (Gen SO1, Configuration 8)

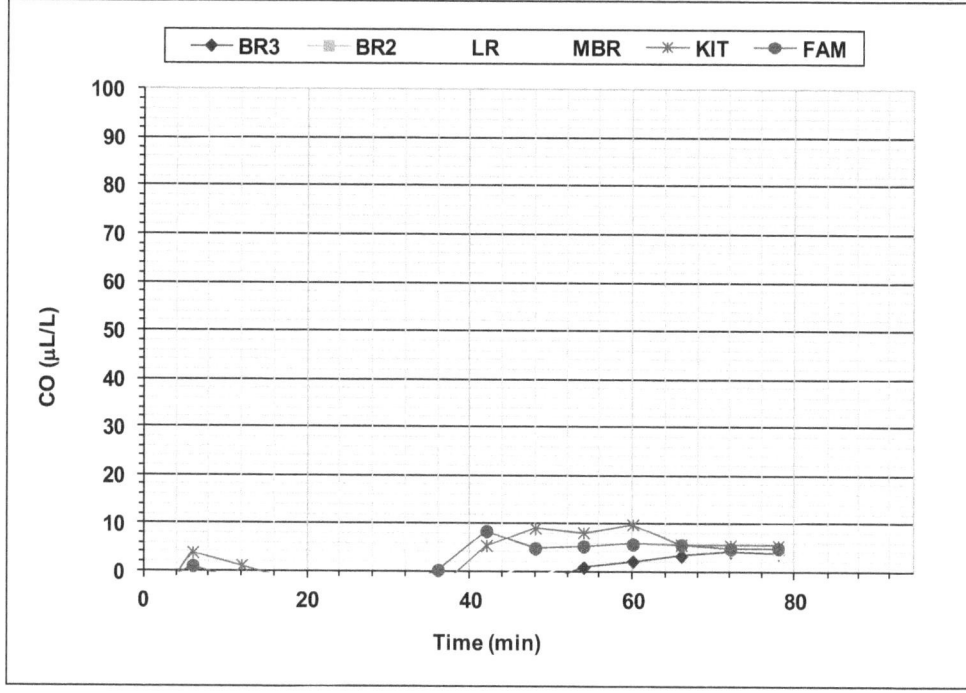

www.ingramcontent.com/pod-product-compliance
Lightning Source LLC
Chambersburg PA
CBHW080251180526
45167CB00006B/2491